EAI/Springer Innovations in Communication and Computing

Series editor

Imrich Chlamtac, European Alliance for Innovation, Ghent, Belgium

Editor's Note

The impact of information technologies is creating a new world yet not fully understood. The extent and speed of economic, life style and social changes already perceived in everyday life is hard to estimate without understanding the technological driving forces behind it. This series presents contributed volumes featuring the latest research and development in the various information engineering technologies that play a key role in this process.

The range of topics, focusing primarily on communications and computing engineering include, but are not limited to, wireless networks; mobile communication; design and learning; gaming; interaction; e-health and pervasive healthcare; energy management; smart grids; internet of things; cognitive radio networks; computation; cloud computing; ubiquitous connectivity, and in mode general smart living, smart cities, Internet of Things and more. The series publishes a combination of expanded papers selected from hosted and sponsored European Alliance for Innovation (EAI) conferences that present cutting edge, global research as well as provide new perspectives on traditional related engineering fields. This content, complemented with open calls for contribution of book titles and individual chapters, together maintain Springer's and EAI's high standards of academic excellence. The audience for the books consists of researchers, industry professionals, advanced level students as well as practitioners in related fields of activity include information and communication specialists, security experts, economists, urban planners, doctors, and in general representatives in all those walks of life affected ad contributing to the information revolution.

About EAI

EAI is a grassroots member organization initiated through cooperation between businesses, public, private and government organizations to address the global challenges of Europe's future competitiveness and link the European Research community with its counterparts around the globe. EAI reaches out to hundreds of thousands of individual subscribers on all continents and collaborates with an institutional member base including Fortune 500 companies, government organizations, and educational institutions, provide a free research and innovation platform.

Through its open free membership model EAI promotes a new research and innovation culture based on collaboration, connectivity and recognition of excellence by community.

More information about this series at http://www.springer.com/series/15427

Syed Abdul Rehman Khan

The Critical Success Factors of Green Supply Chain Management in Emerging Economies

Springer

EAI
RESEARCH MEETS INNOVATION

Syed Abdul Rehman Khan
Tsinghua University
Beijing, China

ISSN 2522-8595 ISSN 2522-8609 (electronic)
EAI/Springer Innovations in Communication and Computing
ISBN 978-3-030-42741-2 ISBN 978-3-030-42742-9 (eBook)
https://doi.org/10.1007/978-3-030-42742-9

This Springer imprint is published by the registered company Springer Nature Switzerland AG
The registered company address is: Gewerbestrasse 11, 6330 Cham, Switzerland

The author dedicates this book to his parents:
Syed Hameed Khan
Rasool Syeda

Preface

Society has started to realize that living unsustainably is not an option for the human race. The Anthropocene is upon us, where man's actions can cause irreversible changes to our air, land, and water environment. We can alter the climate of the entire Earth with our actions. We are capable of causing entire species to extinction, never to be seen on Earth again.

These anthropocentric effects are due to our social and economic behaviors. We are currently, as a world, consuming more than one Earth can handle. If the global consumption rate were the same as some of our most developed regions, it would require up to a dozen Earth's worth of resources.

The detriments to the environment and natural resources caused by our activities can be traced to businesses and industries. They have the greatest environmental impact, all in order to serve our needs. It is the processes and materials they use, how they design our products and services, and how products are stewarded by these companies.

It is only recently, that greening our businesses and their supply chains has come to the forefront. The topic includes thousands of studies and publications from across disciplines considering how to make our industry and economies greener.

There is no doubt that manufacturing has been one of the critical areas that support and influence a nation's economy since the eighteenth century. As the primary driving force behind economic growth, manufacturing serves as the foundation of and contributes to other industries. In the past, manufacturing has contributed significantly to modern civilization and created the momentum that drives today's economy. However, due to the substantial manufacturing activities, particularly in emerging economies, environmental sustainability is on risk, which created a challenge for governmental bodies and policymakers to prevent harmful activities before it is too late.

As a general overview, Chap. 1 begins with a clear definition of green supply chain management and their needs in today's businesses, specifically in emerging countries. Cchapter 2 is based on in-depth literature reviews and hypotheses development. Chapter 3 covers the theoretical framework and methodology adopted to examine the critical success factors and barriers faced by firms. Chapter 4 includes different statistical modeling, statistical techniques, and results of the study. Chapter 5 contains a detailed discussion on the study results and comparison with previously published literature. Finally, Chap. 6 covers concluding remarks and policy implications to the managerial staff and governmental bodies for the sake of better environmental sustainability.

Beijing, China
December 6, 2019

Syed Abdul Rehman Khan, PhD

Acknowledgments

First of all, the author would like to disclose that the book is an extended version of his Ph.D. thesis. Second, the author loves to thank the famous Sufi Scholar Sahibzada Asim Maharvi for his encouragement and support throughout the project. Finally, the author thanks his lovely parents, whose care and prayers for his success have seen this project to fulfillment.

"This book is partially supported by the China Postdoctoral Science Foundation (No. 2019M660700) and the Beijing Key Laboratory of Megaregions Sustainable Development Modelling, Capital University of Economics and Business (No. MCR2019QN09)."

Contents

About the Author

Syed Abdul Rehman Khan is a teacher of Supply Chain and Logistics Management. He obtained his Certified Supply Chain Professional (CSCP) certificate from the United States and completed his Ph.D. in China.

Since 2018, he has been affiliated with Tsinghua University as a postdoctoral researcher. He has more than 9 years of core experience of supply chain and logistics at industry and academic levels. He has attended several international conferences and also has been invited as a keynote speaker in different countries. He has published more than 60+ scientific research papers in different well-renowned international journals and conferences. He is a regular contributor to conferences and workshops around the world. Besides, he has achieved scientific innovation awards three times, consecutively, from the Education Department of Shaanxi Provincial Government, China.

Also, he holds memberships in the following well-renowned institutions and supply chain bodies/associations: APCIS, USA; Production and Operations Management Society, India; Council of Supply Chain Management Professionals, USA; Supply Chain Association of Pakistan; and China Society of Logistics.

Chapter 1
Introduction to the Green Supply Chain Management

1.1 Background

In many countries, public and private organizations are procuring green products, and they are also providing advantages to the environment and companies as a competitive edge over other organizations that are not working on sustainability. There public awareness of environmental issues is immensely increased. Moreover, in today's world, consumers are demanding green products because now, consumers are more conscious regarding environmental issues (Khan & Zhang, 2019). Furthermore, the pressure from the consumer's side insists on the organization to produce environmentally friendly products. There is no other option for companies except producing greener products and switching their supply chain toward green and environmentally friendly supply chains, because companies manufacture products for consumers, and finally, consumers decide whether they buy the products or not. As we have discussed, consumers are more conscious about their health and his/her family health and also about the environment. In simple words, consumers generate money for companies in terms of profits.

The former chairman of the UK Institute of Logistics, Jonathan Weeks, defined logistics as the one circle of materials, component movements from the Earth to manufacturing, warehousing and distribution, sale to consumers, and after consumption going back to the Earth (Khan, 2019; Khan, Dong, & Zhang, 2018; Khan, Golpira, & Zhang, 2018; Khan, Zhang, Anees, et al., 2018). Many companies consider reverse logistics as a drain over profits. However, in today's dynamic and competitive market, it can be taken as a tool for customer satisfaction. Nowadays, many firms are thinking of building reverse logistics in their supply chain system. There is no doubt that reverse logistics is a small part of the environmentally friendly supply chain and the first step toward greener products.

On the other hand, this first step (reverse logistics) is significantly crucial without the successful implementation of reverse logistics; companies are unable to go for

S. A. R. Khan, *The Critical Success Factors of Green Supply Chain Management in Emerging Economies*, EAI/Springer Innovations in Communication and Computing,
https://doi.org/10.1007/978-3-030-42742-9_1

the second step but produce environmentally friendly products. Besides, today's market is the buyer market, not a seller market. So companies need to be customer-centric and fulfill the demands of consumers (Khan, Qianli, et al. 2017; Khan, Dong, Zhang, & Khan, 2017).

These are some significant reasons to produce environmentally friendly products:

- Public awareness about the environment pollution
- Tough competition
- Government strict laws for waste disposal
- Recycling and remanufacturing products to save money

1.1.1 Public Awareness About the Environmental Impact

The serious problems of environmental pollution growing public awareness campaigns. Besides, the government is also using several channels for public awareness, including print media, electronic media, adult education, and college education. The movement of the green environment is proliferating in European countries; on the other hand, Asian countries are following them.

Since the last couple of decades, several research papers have been published in the field of environmental studies, and most studies are academically based on providing precocious awareness to the public. On the other hand, a limited number of research are published on practical implications to help firms in environmental policy making (Taylor, Marshall, Mccarthy, Heavey, & Mcgrath, 2014). In the perspective of Pakistan, several NGOs (nongovernment organizations) are working to promote environmental sustainability and also creating pressure on firms to adopt green practices in their business operations. Pakistan is an agricultural country, and 70% of GDP (gross domestic product) shares come from agricultural products. In simple words, Pakistan needs strong environmental policies to protect environmental sustainability and prevent climate change. The environmental issues are creating many problems in big cities of Pakistan, such as Karachi, Lahore, and Peshawar. As per a study, the government is not taking extreme initiatives to protect environmental sustainability, but on the other side, customers are well aware of pollution consequences through the Internet and international media.

One key reason for environmental degradation is industrialization; the government alone cannot protect environmental sustainability; public awareness plays a significant role in controlling pollution and climate change issues (Meneghetti & Monti, 2014). Environmental education is essential to encourage customers and firms to take environmental friendly initiatives in their society. It sensitizes the society and interested individuals regarding environmental challenges and issues to build expertise and skills, thereby offering suitable solutions (Khan, Qianli, et al., 2017). Loss of biodiversity, ozone layer depletion, climate change, declining fisheries, illegal trade of endangered species, land degradation, destruction of habitats, depletion groundwater supplies, environmental pollution, stormwater, and sewage disposal pose a severe threat to green systems in urban, forest, rural, and marine ecosystems. Undeniably, informal and formal education about the environment will give the interested individuals the knowledge, skills, values, and tools needed to solve the environmental challenges on a domestic and international level (Kim & Min, 2011).

1.1.2 Tough Firm Competition

In the last couple of decades, the green supply chain market is growing speedily all around the globe. Many countries started to green their business operations, and customers are also well aware of environmental sustainability. Due to the higher pressure of government and customers, firms are shifting their polluted supply chain over green and clean supply chains (Gruner & Power, 2017). On the other hand, many developed countries, including the USA, Germany, France, and the UK, already adopted green practices in their supply chain activities to protect environmental sustainability and reduce harmful effects on human lives (Shin and Thai, 2016).

1.1.3 Government Strict Laws for Green Performance

The laws of the environment also are called the natural resources law; the term is describing the network of statutes, treaties, regulations, and customary laws addressing the impact of human activities on environmental sustainability. The fundamental law of environment regimes addresses environmental pollution. An associated but different set of regulatory regimes, now strictly influenced by environmental legal philosophies, emphasizes the management of specific resources, including minerals, forests, and fisheries.

1.1.3.1 Pollution Control

The quality of air has been destroyed due to heavy pollution in the environment. The laws of air quality are usually explicitly designed to protect people's health by eliminating airborne. Other initiatives are designed to address broader green issues like limiting carbon emissions to protect the ozone layer and preventing climate change and acid rain (Taylor, Marshall, et al., 2014). The efforts of regulatory authority are identifying and limiting air pollutants and enforcing necessary and suitable mitigation technologies.

The quality of water laws protected sea pollution, including ground, surface, and stored drinking water. Many qualities of water laws such as drinking water regulations can be designed concerning people's health. On the other hand, chemicals and massive waste go into the sea and river, which not only create destructive situations for human lives but create severe consequences for flora and fauna lives.

The critical efforts of regulatory authorities are identifying and categorizing water pollutants and limiting pollutant discharges from effluent sources (Jakowczyk, Quariguasi, Neto, Gibson, & Van, 2017). The laws of waste management, the treatment of wastage, storage and disposal of all kinds of waste, including hazardous waste, solid waste, and nuclear waste. Generally, the laws of waste are designed to reduce the uncontrolled dispersal of waste materials into the atmosphere in a way that may cause environmental problems and pollutions, and laws also encourage recycling, remanufacturing, and reducing approaches to promote green supply chain management and green practices in business (Lee & Wu, 2014). The efforts of the government include identifying and classifying the types of waste and their treatment and disposal practices.

Environmental pollution destroys sustainability and creates significant problems for human lives. The laws of pollution control are designed to solve the issues of the environment and consequently must often explain not only the necessary response actions but also the need of parties to take initiatives to keep the environment clean and green (Meneghetti & Monti, 2014). The requirement of regulatory authorities may include liability allocation, emergency response, feasibility studies, site assessment, and post-remedial monitoring and remedial investigation. Undeniably, the chemical used in the products severely damages environmental sustainability and human lives, while flora and fauna are also badly destroyed due to industrialization. The concept of the green supply chain not only protect environmental sustainability but also protect the lives of humans (Liu, Tang, & Xue, 2012). The government efforts to protect the environment include banning hazardous chemicals, including polypropylene and black plastic bags.

1.1.3.2 Resource Sustainability

Impact Assessment

The impact of environmental assessment is the assessment of consequences which means a positive or negative impact on the environment. EIA (environmental impact assessment) is usually used when applied to real projects by firms or individuals, and the term "strategic environmental assessment" is applied to plans, policies, and programs.

Water and Mineral Resources

The resources of water and mineral are essential factors for country development and growth. The mining is affected by many regulations regarding the safety and health of miners, as well as the environmental impact of carbon emissions, GHG (greenhouse gases), and global warming.

Forest Resources

The resources of the forest protection environmental sustainability and keep the air quality good for human lives. However, unfortunately, the forest is disappearing due to industrialization, and carbon emissions are also increasing in dangerous positions. Forest absorbed air pollution (carbon dioxide) and emitted oxygen, which protects the human, flora, and fauna lives. Mainly it is the government and regulatory authorities work to protect trees and to plant trees, but due to greater public awareness, different kinds of NGOs and individual people are also planting trees (Ye, Zhao, Prahinski, & Li, 2013).

1.1.4 Recycling and Remanufacturing Products to Save Money

The process of recycling is to convert waste materials into a useable form and shape. The recycling process and remanufacturing process have an important step in the concept of GSCM ("green supply chain management"). Waste disposal can reduce the cost of products and also increase environmental sustainability. Furthermore, the recycling idea is vastly used in several countries, including Pakistan, the USA, the UK, and Germany, to control climate change, air pollution, and water pollution. The process of recycling has some ISO standards such as ISO 15270:2008 for the waste of plastic and ISO 14001: 2004 for environmental management control (Brandenburg, 2015).

Recycling may include many kinds of items, such as cardboard, tires, plastic, paper, glass, and electronic items, and each item has their recycling process. The primary procedure of recycling includes collection, sorting, cleaning, and remanufacturing. The two primary types of recycling are recycling of material and producing a new product; for example, using plastic bottles and converting them into different shapes. However, this is often difficult or too expensive (compared with manufacturing the same item from raw materials/components), so mainly (Savita, 2014) recycling of several products involves their reuse in manufacturing different materials or due to their hazardous nature such as thermometers.

1.2 International Laws

Environmental pollution problems have been raised at various levels, and several studies have been published on the subject. The international law has become a multitude of international agreements between different countries to control global warming, carbon emissions, and the impact of greenhouse gases on human life and environmental sustainability. Common law is the source of international environmental law. These are the rules and norms that various countries use as a habit and are very widespread because they interfere with individual countries. When a rule becomes a customary law, it is not very clear and understandable, and then various arguments from different countries appear. An example of customary law is to quickly warn other countries about icons of environmental damage to which another country may be exposed (Keehung Lai & Wong, 2012). Several legally binding agreements are covered by international agreements, which include pollution of the sea and the atmosphere through biodiversity and the protection of wild animals. Agreements and international organizations include IUCN (International Union for Conservation of Nature), CITES (Convention on International Trade in Endangered Species of Wild Fauna and Flora), and the World Commission on Environment and Development.

International environmental law includes the opinions of international courts and tribunals, although they have limited authority. One of the critical challenges in international decisions is to examine a satisfactory compensation for environmental losses.

1.2.1 European Union Countries

The environmental issues in the European Union are valid throughout the EU and various directives that should be executed into national legislation from 28 different member countries. The main focus of EU states are on the following environmental issues:

- Climate change
- Water protection and management
- Soil protection
- Noise pollution
- Air pollution
- Waste management
- Civil protection
- Cooperation for the environment with other than European Union countries

1.2.2 Asian Countries

The environmental situation in Asian countries is also not very good, and due to these reasons, many Asian countries built a union called AECEN (Asian Environmental Compliance and Enforcement Network) (Kim & Min, 2011). In this union, a total of 16 Asian countries are actively playing roles such as Pakistan, China, Cambodia, Indonesia, Maldives, India, Japan, Malaysia, Korea, the Philippines, Nepal, Sri Lanka, Singapore, Vietnam, Laos PDR, and Thailand.

Twelve years after the historic Rio Summit on environment and development, countries around the world are still struggling with complex challenges that reconcile development goals with environmental imperatives. In Southeast Asia, as in other countries, national governments try to address these challenges, including the adoption of new regulations on natural resource management and environmental protection, the ratification of relevant international conventions, and the establishment or strengthening of institutions required in the field of environmental management. In recent years, several new national standards have been introduced in the Southeast Asian region to address the problems of environmental management and natural resources.

1.3 Highlights of Environmental Laws in Pakistan

Industrial activities are defined in PEPA as "any operation or process for the manufacture, production, formulation, synthesis, modification, repair, decoration, finishing, packaging or treatment of any article or substance for its use, sale, transport, supply or utilization or extraction; for pumping water or sewage, or for producing, converting or transmitting energy or for any other industrial or commercial purpose" (section 2 (xxii)). The broad scope of regulation of industrial activity, resulting from the definition, is reflected in the applicable laws, in particular in the ban on specific emissions with the provisions on charging for pollution (section 11), IEE/EIA (section 12), licenses (Sections 14 and 15), and orders for environmental protection (section 16). The project is defined as covering the construction, operation, alteration, extension, repair, dismantling, or abandonment of factories or other devices (sections 2 (xxxv) (c) and (xxxv) (f)). PEPA requires that parties who want to start a project send an EIA or IEE to the Federal Agency (section 12). Meanwhile, the Federal Agency may issue an environmental protection order in cases where discharges, emissions, disposal, or handling of hazardous substances may hurt the environment.

The penalty for non-compliance with the provisions on discharges and emissions (section 11) and environmental protection orders (section 16) is a fine that can be extended to one million rupees for the first offense (§ 17 paragraph 1). In the event of non-compliance with regulations on the handling of hazardous substances 6.1.3.7 Mining Law, 1923 6.1.3.8 Land Acquisition Law (1848) 6.1.4.1 Pakistani Environmental Protection Law, 1997 6.1.4 Non-extraction Industry 53 Environmental Law in Pakistan Federal 06 Processes and institutions (section 14), fines can be extend to 100,000 rupees (Section 17 (2)). In addition to imprisonment for a maximum of 2 years, penalties for repeated offenders may include closure or confiscation of factories, machinery, and equipment involved in the crime in order to restore the environment at the expense of the perpetrator and in order to pay compensation for any loss, injury, or damage to health or property caused by the violation (section 17(5)).

Under these rules, production and processing projects must be subjected to prior IEE or EIA. The small commercial industry and operations requiring a prior IEE are listed in Section C of Annex I. More advanced manufacturing and processing operations which require prior environmental assessment are listed in Annex II, Section B.

This Act establishes the Heavy Industry Council, which performs commercial and administrative functions concerning Heavy Industries Taxila, including massive factories, as well as other allied facilities belonging to the federal government. The Act defines the corporate existence and functioning of the board (section 3). The federal government is empowered to set rules for achieving the objectives of this law (chapter 14). The Council is authorized to delegate its powers to one of its members (section 11). The management manages only massive Taxila reconstruction factories. This Act establishes the Pakistani Standard and the Quality Control Office, which oversees standardization and quality control in the industrial products and processed products sector and conducts product tests for import, export, and local consumption. The Office is responsible for training programs, classification and labeling products, and communication with national and international public and government institutions in matters related to standardization (section 8). The law does not refer to international standards, in particular ISO 14000. The federal government may establish rules to achieve the objectives of this Act (section 26).

This Act regulates employment, working hours, working conditions, and facilities that will be provided at the workplace. The Act mainly concerns the issue of industrial relations but includes a detailed chapter on the health and safety of employees (Chapter 3, Sections 13–33). The law requires factories to be kept clean, and "spills from any drain, toilet, or other nuisance" are removed at regular intervals (section 13). Te and 6.1.4.2 Review of the initial environmental review and assessment of environmental impact 2000 (SRO 339 (I) / 2000 of 13 June 2000) 6.1.4.3 Council Act on Heavy Industries of 1997. 6.1.4.4 Normative Authority and Quality control of Pakistani law 1996 6.1.4.5 Plant Law 1934 54 Environmental Protection Law in Pakistan Federal 06 Processes and Institutions other health and safety regulations apply only to workers in factories. The law does not mention the elimination of waste generated during the production process, nor does it require measures to mitigate the impact of such waste on the environment. All rights resulting from this act, including the power to set standards, were transferred to the provincial governments.

Since the problem of environmental pollution and ecology is a concurrent legislative list of the constitution, both federal and provincial governments can enact laws regulating natural resources. Constitutional weapons divide the jurisprudence of the properties of matter in general, but not in the specific context of natural resource management. In critical areas of the country, including in the tribal and northern areas, federal and provincial legislations do not apply automatically, unless the government with jurisdiction over the area expands the application of the law in this area. For example, the Forest Act does not apply to FATA. With few exceptions, federal laws on natural resources and other delegated executive ordinance or provincial governments or agencies and bodies are created explicitly for this purpose. Case law explains that the delegated authority is not final and can return to the posting authority. None of the cases explicitly refers to the transfer of power in the field of natural resource management. Even when delegated powers, gaps in business rules in the jurisdiction of ministries and the functions of its units mean that there is no operational basis for integrated management of natural resources. The Pakistani Environmental Protection Board and the Pakistan Environmental Protection Agency can create a framework for integrated management, as part of Pakistan Environmental Protection (PEP) broad interpretation.

Billion Tree Projects in Pakistan

> The project started in 2013 by the central and provincial governments of Pakistan, which has been successfully ongoing, and due to this project, environmental sustainability significantly has been improved. On the other hand, several international organizations, institutions, and NGOs appreciated the Pakistan government project. The following are the critical remarks from a couple of international organizations.

The Report of the UN (United Nation 2016)

> This is a project for the future of Pakistan. It is providing a green, breathable environment and green jobs but is also building up Pakistan's much-needed defense against the high climate vulnerability that it faces. On the other hand, it will also reduce global warming.

The Report of WWF (World Wildlife Fund 2017)

> It significantly contributes to the global efforts for sequestering carbon and mitigating climate change.

The Report IUCN (International Union for Conservation of Nature 2017)

> Congratulates the Pakistani Government on reaching this momentous milestone. The Billion Tree Tsunami initiative is a true conservation success story, one that further demonstrates Pakistan's leadership role in the international restoration effort and continued commitment to the Bonn Challenge.

1.4 Definition of GSCM in the Context of Pakistan

Definition of GSCM (green supply chain management) or supply chain management with respect for the environment: conventional SC (supply chain) to add activities aimed at reducing the environmental impact of the product in all its PLC aspects (product life cycle), e.g., eco-design, reduction of materials/harmful ingredients, and recycling of products. Green supply chain management is a combination of regular supply chain management and consumer awareness of environmental issues with the emphasis on the production of organic products and reducing the harmful effects on the environment at each of the stage of the value chain through:

1. Reducing the usage of energy
2. Reducing the usage of natural resources
3. Reducing pollution-related issues
4. Increasing reuse, refilling, remanufacturing, and recycling to harness the again use of components and raw materials

The idea of managing the ecological supply chain is to integrate ecological thinking with the supply chain and companies. GSCM has a significant impact on the environmental impact of companies that participate in supply chain activities and contribute to improving sustainability performance. In Pakistan, an environmentally friendly supply chain is becoming more cost-effective, and some companies are changing their supply chain to an environmentally friendly supply chain. The most important question, however, is why companies are changing toward environmentally friendly supply chain. So the answer to this question is not too complicated because many factors put pressure on the administration to implement an environmentally friendly supply chain.

Undeniably, the green supply chain management process can vary from one company to another and from one product to another. However, the ending of the green supply chain procedure is the same. The green supply chain management not only saves the environment. Nevertheless, it also provides an opportunity for companies to grasp the competitive edge in their market and increase their profitability (Figs. 1.1 and 1.2).

1.5 The Network of Reverse Logistics

The network of reverse logistics use for multiple purposes including:

- Remanufacturing
- Refilling
- Reuse
- Repair
- Recycling
- Recover energy or disposal

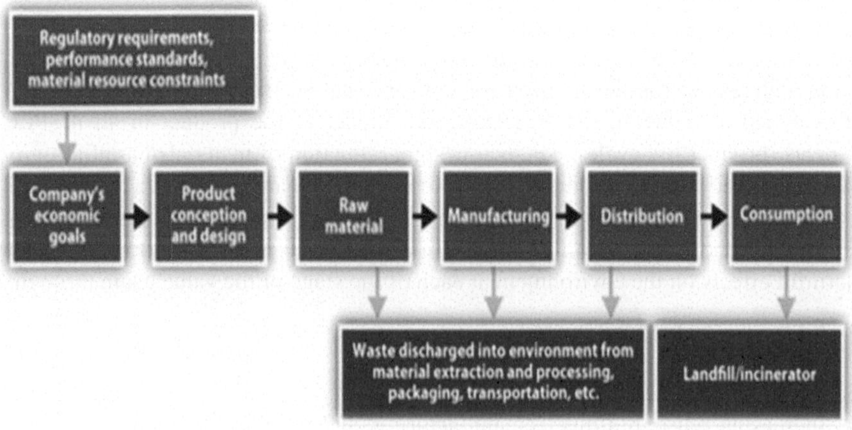

Fig. 1.1 The conventional supply chain

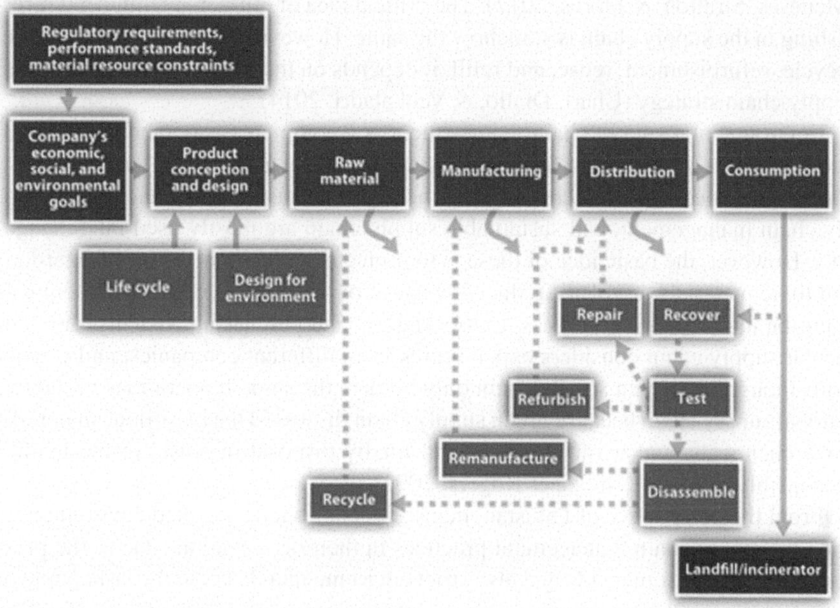

Fig. 1.2 The green supply chain

It mainly depends on the nature of the product if companies can remanufacture, reuse, or refill. Undoubtedly, the regeneration process is different from other recovery processes after use, such as reuse, regeneration, repair, and recycling. Regenerated products are subjected to an extensive production process as compared to renovation or recycling, and therefore provide better results in terms of quality (Yusop, Wahab, & Saibani, 2016). On the other hand, the recycling process is not the same as in the case of regeneration products, and for this reason, the quality also differs. Current production practice in which OEM (original equipment manufacturer) plays an important role is regeneration which means that the used item is involved in the donating process to the minimal original equipment manufacturer in order to maintain the quality and specification of the products. According to the British Standards Institution, the definition of regeneration is "restoring the product at the end of its useful life to at least its actual performance with a higher or lower equivalent quality of the newly created product" (British Standard Institution, 2010). It estimated that the global market for reproduced vehicle parts would reach USD 122.8 billion a year in 2018 (Global Industry Analysts, 2013). Steinhilper (2001) stated that regenerative processes are widely used in the automotive sector and account for 2/3 of the entire regeneration industry.

In the supply chain, refilling means used products after the shelf life, such as plastic containers, glass bottles, and metal cylinders, are refilled. The concept of refill is not very different than reuse, and the main objective is the same – to reuse the products. In short, the word "refill" is usually used for bottles and containers

(Meneses, Stratton, & Flores, 2017). The critical idea of reuse, recycling, and refurbishing in the supply chain is somehow the same. However, when firms need to use, recycle, refurbishment, reuse, and refill, it depends on the product nature and firms' supply chain strategy (Chari, Diallo, & Venkatadri, 2014).

Environmental supply chain management is a complete phenomenon and practice to save the environment from negative and harmful activities of the businesses. There is no doubt that the terms environmentally friendly supply chain, green supply chain management, and sustainable supply chain are mostly used interchangeably. However, the basic idea of these is to protect the environment from hazardous and toxic materials or at least reduce the usage of harmful materials and chemicals from the manufacturing process, transportation, and purchasing. Usually, environmental supply chain considers expenditures from different companies, and usually firms thought the green supply chain only hinders the smooth operations of the traditional supply chain because green supply chain discussed the hazardous materials, toxic chemicals, and waste products and finally disposal of waste in the landfill responsibly (Tibben-Lembke & Rogers, 2002).

From the perspective of Pakistan, many companies have adopted environmental friendly supply chain management practices in their organizations due to the pressure from the consumer side and also strict environmental laws. In the last couple of decades, almost 2600 companies fully adopted green supply chain practices in their manufacturing process, and these numbers are continuously growing. Besides, companies are not only implementing green supply chains from the fear of strict environmental policies but also for creating a competitive edge in their industry.

1.6 Differences Between Forwarding and Reverse Logistics

Reverse logistics is a small part of green supply chain management. Companies cannot be called a green supply chain without the implementation of reverse logistics on the first step (Tibben-Lembke & Rogers, 2002). Reverse logistics refers to the critical role of logistics in goods returns, recycling, reuse of materials, source reduction, materials substitution, refurbishing, waste disposal, remanufacturing, and repair; in the context of engineering logistics, it is referred to as RLM (reverse logistics management) and is a business model which applies best logistics management and engineering methodologies across the companies (Fig. 1.3).

The forwarding logistics is defined as the products, raw materials, and components that move from upstream (supplier/manufacturer) to downstream (customer/consumer), and information and cash that travel from downstream to upstream. In simple words, forward logistics only delivered products to the downstream layers.

In light of are-Lembke, many companies use centralized distribution centers to process return components or goods. From the theoretical perspective, the forward distribution centers can also be used for return products, but as they explained, there is much temptation for RL (reverse logistics). When the goods are received at centralized distribution centers, a thorough determination should be completed regarding where the goods should be dispatched. Usually, employees at distribution centers assess the condition of every incoming product and judge the best location to dispose of the product. In the research of Thierry et al. (1995), probable destinations for goods are:

- Reuse
- Sell to the broker
- Return to the supplier
- Repackage
- Recycle
- Refurbish or remanufacture
- Donate to charity
- Landfill

The possible destination of returned products from customers is mentioned above (Thierry et al., 1995). The difference between forwarding logistics flow of information and reverse logistics flow of information can easily be seen in Figs. 1.4 and 1.5.

Several researchers have discussed that planning for forwarding distribution is more straightforward than planning for reverse logistics due to much involvement of uncertainty in reverse logistics. In remanufacturing, planning and forecast are not only required from customer demand, which is always challenging, but also from the availability of goods to be manufactured (Table 1.1).

In the context of retail, planning and forecasting for RL are made complex due to individual customers/consumers. However, somehow, the general trend may be observed. When scheduling promotional activities, marketing and sales department already closely work with forwarding logistics to confirm and ensure that the customers are satisfied with products. Usually, large numbers of sales also create or show some trend of return goods. Usually, in promotional activities, salespeople do over-promise with customers and try to use up-selling techniques, which results in a higher volume of returned products after promotional activities. Besides, most

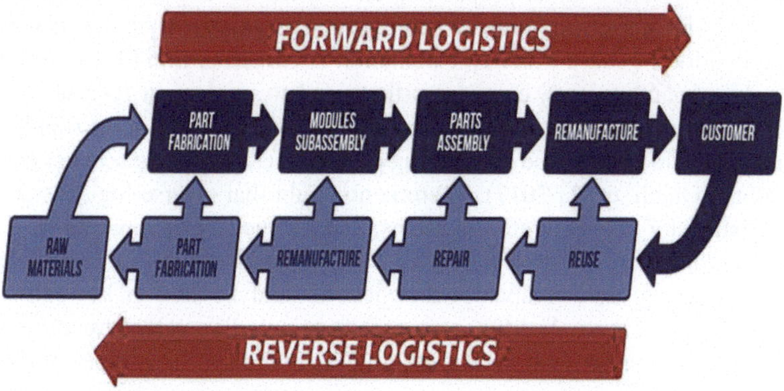

Fig. 1.3 Forward and reverse logistics

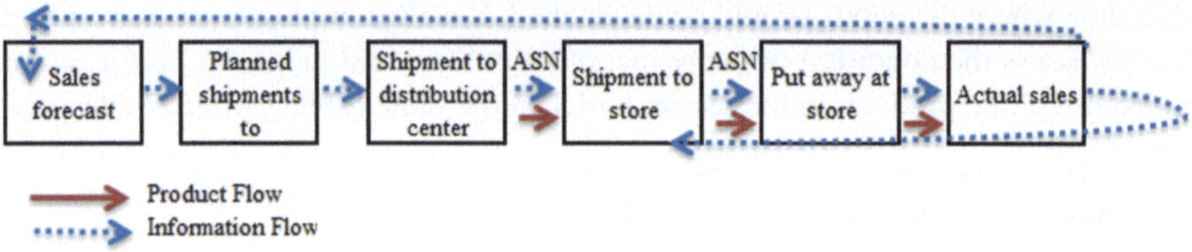

Fig. 1.4 Forward logistics flow of information for retail

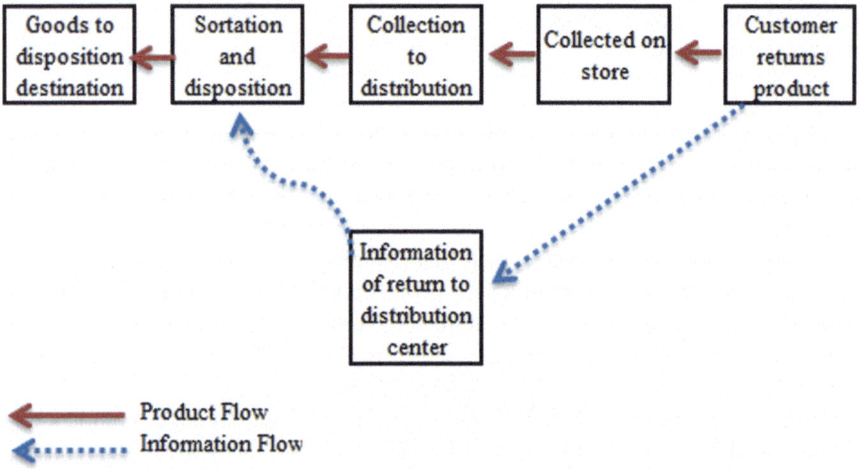

Fig. 1.5 Reverse logistics flow of information for retail

of the time due to a higher volume of returned products, the company's reputation and image also become distorted in front of consumers (Khan, Zaman, & Zhang, 2016).

According to several researchers, reverse logistics is a small part of the environmental supply chain; without proper implementation of reverse logistics, it is not possible to implement GSCM (green supply chain management) in the business. So it means that for the successful implementation of GSCM or ESCM (environmental supply chain management), companies first need to implement reverse logistics. Besides, the involvement of suppliers, customers, and all entities of the supply chain is also essential to be involved in greening processes (Khan, Dong, Zhang, & Khan, 2017; Khan, Qianli, et al., 2017). There is no doubt that reverse logistics and forward logistics are two different directions, and for the implementation of reverse logistics, companies need to be very careful (Lai and Wong. 2012). For the execution of reverse logistics, companies first need to analyze their end-to-end supply chain and business through different approaches including value stream mapping and total quality management. In the last couple of decades, many types of research

Table 1.1 Differences in reverse logistics and forward logistics

Reverse logistics	Forward logistics
Forecasting more complex	Forecasting relatively easy or straightforward
Many to one transportation	One too many transportations
Goods' quality are not uniform	Goods' quality are uniform
Packaging of products are usually damaged	Packaging of products are uniform
Destination unclear	Destination well clear
Exception driven	Standardized channel
Disposition not well clear	Disposition options well clear
Pricing depends on many elements	Pricing relatively uniform
Speed usually not considered a priority	Speed considered a priority
Reverse costs are less visible (directly)	Forward distribution costs tightly controlled by accounting
Management of inventory is not consistent	Management of inventory consistent
PLC (product life cycle) very complex	PLC manageable
Negotiation difficult by additional considerations	Negotiation between parties clear or straightforward
Marketing complicated by many elements	Marketing methods well-defined
Visibility of process less transparent	Real-time information sharing, easily tracked products

have been done over reverse logistics and environmentally friendly supply chains, and most studies are conducted in the Western world. Reverse logistics or reverse flow of materials/components/products will finally be reuse or remanufactured (Lorenzoni and Lipparini, 1999). In the forward logistics usually, forecast to the future sales and then products send to the DC (distribution centers), and finally, goods arrived at retail outlets for final customers.

However, on the other hand, reverse logistics has much less visibility and coordination among supply chain entities, and it is reactive. In reverse logistics, companies do not initiate activity as a result of planning and decision, but in response to actions through downstream channels (consumers).

On the ground reality, it is challenging to control reverse logistics activities without the involvement of downstream echelons, including consumers, retailers, and distributors. Most of the time, consumers return products to a retailer for repairing, remanufacturing, or recycling. Nevertheless, the retailer keeps these products/items for a week or more, making them a consolidated shipment for sending to the distributor, and the distributor also takes time for making it more significant consolidated shipment and finally sends it to the manufacturers, and these all procedures sometimes take a month or more. Besides, with little or no visibility among supply chain partners (retailer, distributor, manufacturer, and suppliers), this procedure becomes more complex and painful for final customers or consumers.

1.7 Pakistani Manufacturing Sector

The manufacturing industry of Pakistan is the second largest industry in Pakistan, and this industry contribution in GDP (gross domestic product) is around 15%. Besides, the manufacturing industry provides job opportunity of 16% to the total labor force. The manufacturing sector mainly includes textile, engineering goods, agro-based industry, chemical industry, and small- and medium-sized enterprises. In the manufacturing industry, LS (large-scale) manufacturing companies share is almost 13% in the GDP, and the remaining are the SS (small-scale) manufacturing companies/enterprises (World Bank, 2015). This sector (manufacturing industry) mainly depends on the utility services, right environment, and policies of the government, FDI (foreign direct investment), credit to the private sector, etc. The major issues with this sector are power shortages due to electricity issues manufacturing firms still not growing as per expectations. Pakistan has many coal mines which can produce electricity. Nevertheless, sustainable development is more important for the Pakistan government as compared to polluted development and an unhealthy environment (Pakistan Bureau of Statistics, 2016; World Bank, 2015).

1.8 Manufacturing Industry Growth

In the last 10 years, the government focused on generating electricity through sustainable methods, including thermal and turbine. The comparison growth of the manufacturing industry is presented in Figs. 1.6 and 1.7. The manufacturing industry of Pakistan is continuously growing, and due to new opportunities for domestic and foreign investors, companies are willing to invest in the manufacturing sector. If we analyze the fact and figure of manufacturing industry according to the PBS

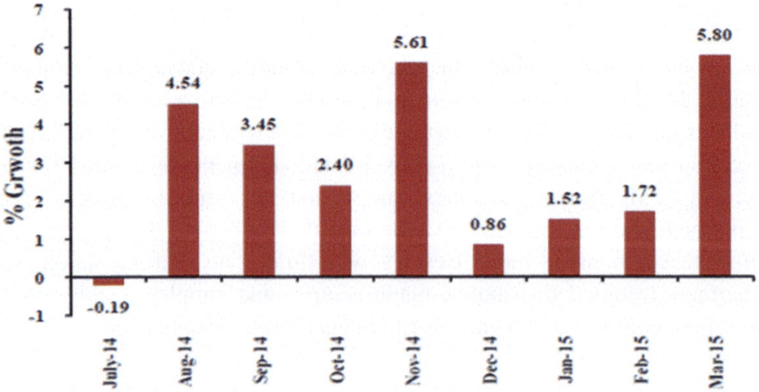

Fig. 1.6 Growth rate of manufacturing industry in 2014–2015

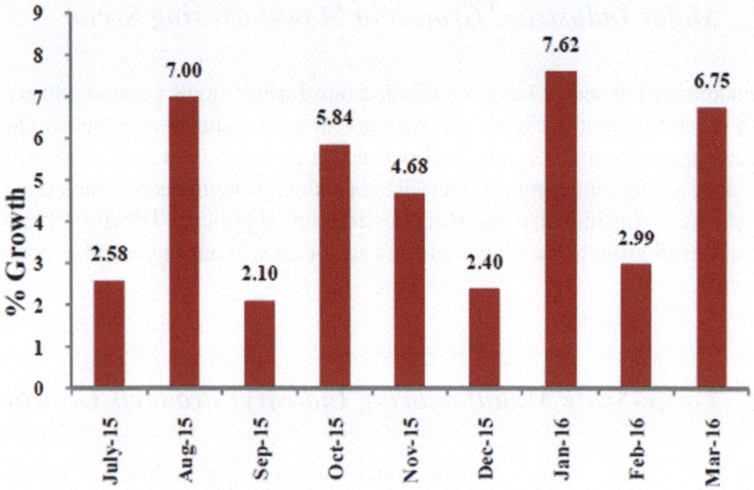

Fig. 1.7 Growth rate of manufacturing industry in 2015–2016

Table 1.2 Group-wise growth change

S.No.	Groups	Weights	% change		% point contribution	
			July–March		July–March	
			2014–2015	2015–2016	2014–2015	2015–2016
1	Textile	20.915	0.97	0.62	0.20	0.13
2	Paper and board	2.314	−5.74	−2.90	−0.13	−0.07
3	Coke and petroleum products	5.514	5.47	2.40	0.30	0.13
4	Rubber products	0.262	1.88	11.68	0.00	0.03
5	Chemicals	1.717	6.67	10.01	0.11	0.17
6	Automobiles	4.613	17.06	23.43	0.79	1.08
7	Beverages, food, and tobacco	12.37	−0.93	3.66	−0.12	0.45
8	Pharmaceuticals	3.620	6.84	7.21	0.25	0.26
9	Electronics	1.963	8.51	−9.98	0.17	−0.20
10	Iron and steel products	5.392	35.63	−7.48	1.92	−0.40
11	Leather products	0.859	9.11	12.18	0.08	0.10
12	Nonmetallic mineral products	5.364	2.71	10.23	0.15	0.55
13	Engineering products	0.400	−10.74	−17.64	−0.04	−0.07
14	Fertilizers	4.441	0.95	15.92	0.04	0.71
15	Wood products	0.588	−78.46	−58.03	−0.46	−0.34

(Pakistan Bureau of Statistics), the manufacturing industry is performing very well as compare to past performance of the manufacturing industry (see Table 1.4) (Pakistan Bureau of Statistics, 2016; Van et al., 2014).

1.8.1 Major Industries' Growth in Manufacturing Sector

The manufacturing sector has been divided into further small groups/industries for microlevel comparison. In Table 1.2, we can realize that almost all groups/industries are growing except textile, code and petroleum products, iron and steel products, electronics, and engineering products. This product growth has declined due to the new policies in this industry and strict environmental policies. However, if we analyze the overall growth, the manufacturing sector growth has increased as compared to last year.

1.8.2 Large-Scale Manufacturing Industry: Products Growth

In Table 1.3, we have further divided products according to large-scale manufacturing industry products and also compared 2016 and 2015 performances of LS manufacturing industry products. There is no doubt that many products of the large-scale manufacturing industry have been increased dramatically as compared to the previous/corresponding year.

From the perspective of Pakistan, as we have discussed, the manufacturing industry also adopted green practices, and still, many companies are switching toward a green supply chain. As we go on further, we need to discuss some more about the manufacturing industry of Pakistan. Whenever we discuss the manufacturing industry, we cannot ignore the textile group of Pakistan. Textile firms cover a significant portion of manufacturing GDP and also provide many opportunities for the labor force. In Fig. 1.8, we can analyze that production is continuously increasing as compared to the corresponding (2015) year. Also, cotton, sugar, and ghee are the backbone of the manufacturing sector in Pakistan. It means that we can judge the growth of the manufacturing sector in Pakistan.

1.9 Production Index of Manufacturing Sector

The production index of the manufacturing sector from 2000 to 2016 is presented in Table 1.4. In this table, we can easily see the continuous growth trend of the Pakistani manufacturing industry. According to the Pakistan Bureau of Statistics, the manufacturing sector is dominant in Pakistan (Nakajima et al., 2017). Further, due to the CPEC (China-Pakistan Economic Corridor), more than hundreds of companies came to Pakistan for investment, and thousands of firms are willing to come in the upcoming years. In the last couple of decades, many foreign companies came to Pakistan to start their operations. Besides the strict policies of sustainable development, many companies have implemented green practices as a whole or partially in their manufacturing operations/processing (Monczka, Handfield, Giunipero, &

Table 1.3 Manufacturing of selected industrial items of LSM

S. no.	Products	Unit	Weight	July-March		% Change (Jul-Mar) 2015–2016	% Point contribution (Jul-Mar) 2015–2016
				2014–2015	2015–2016		
1	Deep freezers	(Nos.)	0.162	56,294	47,987	−14.76	−0.02
2	Phosphatic fertilizer	(N tonnes)	0.400	442,164	499,412	12.95	0.05
3	Refrigerators	(Nos.)	0.239	1,003,795	968,418	−3.52	−0.01
4	Nitrogenous fertilizer	(N tonnes)	4.041	1,940,098	2,256,438	16.31	0.66
5	Cement	(000 tonnes)	5.299	23,459	25,900	10.41	0.55
6	Jeep & cars	(Nos.)	2.818	106,135	137,688	29.73	0.84
7	Upper leather	(000 sq.m.)	0.392	18,371	18,898	2.87	0.01
8	Petroleum products	(000 Liters)	5.410	10,317,669	10,763,101	4.32	0.23
9	Tea blended	(Tonnes)	0.382	92,129	104,871	13.83	0.05
10	Sugar	(Tonnes)	3.545	4,812,408	4,949,653	2.85	0.10
11	Liquids/ syrups	(000 Liters)	1.136	73,424	81,264	10.68	0.12
12	Coke	(Tonnes)	0.104	190,794	57,394	−69.92	−0.07
13	Tablets	(000 Nos)	1.914	19,720,173	20,441,036	3.66	0.07
14	Cooking oil	(Tonnes)	2.227	268,679	291,361	8.44	0.19
15	Cotton yarn	(Tonnes)	12.965	2,513,821	2,552,654	1.54	0.20
16	Cotton cloth	(000 sq.m.)	7.186	776,900	780,233	0.43	0.03
17	Vegetable ghee	(000 tonnes)	1.144	877,467	931,179	6.12	0.07
18	Cigarettes	(Million nos.)	2.125	46,790	42,892	−8.33	−0.18
19	Pig iron	(Tonnes)	1.584	195,741	1,509	−99.23	−1.57

Patterson, 2009). On the other hand, those companies who have still not implemented green practices or sustainable practices in their operations are facing pressure from downstream (customers) side and also facing legal problems from the governmental side (Khan & Qianli, 2017) (Fig.1.9).

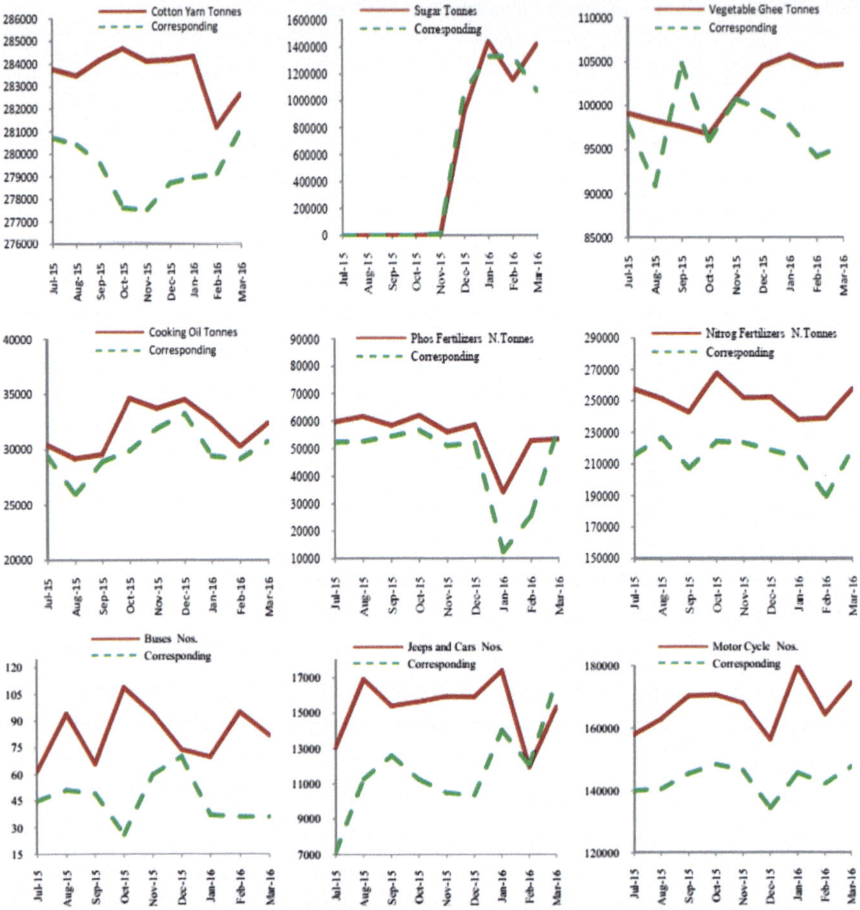

Fig. 1.8 Monthly production trend of goods

1.10 Leading Textile Industry in Pakistan

Pakistan's textiles are number one in Asia. Pakistan is a significant exporter of cotton and textile items in the world. According to the governmental records, Pakistan export from 2008 to 2016 is rapidly increasing, and each year Pakistan exports cotton to the neighboring countries including Iran, India, Maldives, and Nepal. In Table 1.5 (all figures in US$ million), the export volumes during 2015–2016 are the half-year figure (State Bank of Pakistan, 2016; Pakistan Bureau of Statistics, 2016; United Nations, 2016) (Fig. 1.10).

Table 1.4 Production index of manufacturing sector

Year	Production index
2000–2001	111.2
2001–2002	115.2
2002–2003	123.6
2003–2004	126.0
2004–2005	125.0
2005–2006	100.0
2006–2007	109.5
2007–2008	116.1
2008–2009	109.1
2009–2010	109.6
2010–2011	111.1
2011–2012	112.4
2012–2013	117.4
2013–2014	123.7
2014–2015	127.9
2015–2016	133.9

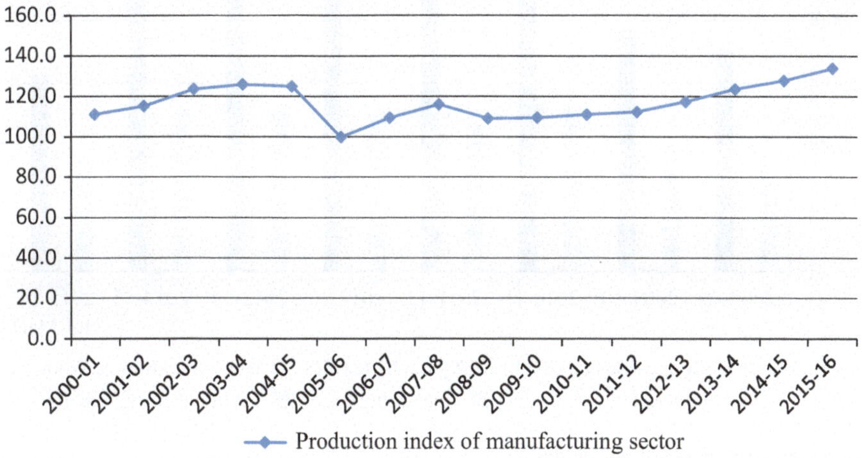

Fig. 1.9 Production index of manufacturing sector

1.11 Automotive Industry

According to the Pakistan Automotive Manufacturers Association, the production of the automotive industry in Pakistan is booming and, as compared to last year's production of vehicles, has increased (Habidin, Fadzlin, Zubir, & Fuzi, 2017). To analyze the facts and figures, we have collected the data of the last 2 years from PAMA (Pakistan Automobile Manufacturers Association). In Table 1.6 (no. of units

Table 1.5 The export of Pakistan textiles

	2008–2009	2009–2010	2010–2011	2011–2012	2012–2013	2013–2014	2014–2015	2015–2016(Jul–Mar)
Cotton textiles and cotton	9308	9754	13,147	11,778	12,652	13,143	12,992	9066
Synthetic textiles	319	446	608	546	406	383	331	222
Woolen textiles and wool	145	137	132	121	122	125	119	74
Total textiles	9772	10,337	13,887	12,445	13,180	13,857	13,590	9438
Total exports	17,782	19,290	24,810	23,624	24,515	25,131	23,885	15,606
Textile as % of exports	55	54	56	53	54	55	57	60

Fig. 1.10 Export of Pakistan textiles

produced), we have mentioned the installed capacity of vehicles, 2014–2015 production, and 2015–2016 production of the vehicle as well as the change in percentage in these 2 years (Fig. 1.11).

Table 1.6 Production of automotive industry

Category	Installed capacity	2014–2015	2015–2016	% change
Cars	240,000	105,267	137,067	30
LCVs	43,900	17,521	29,529	69
Buses	5000	410	746	82
Trucks	28,500	2781	3940	42
Two/three wheelers	2,500,000	811,459	982,423	21

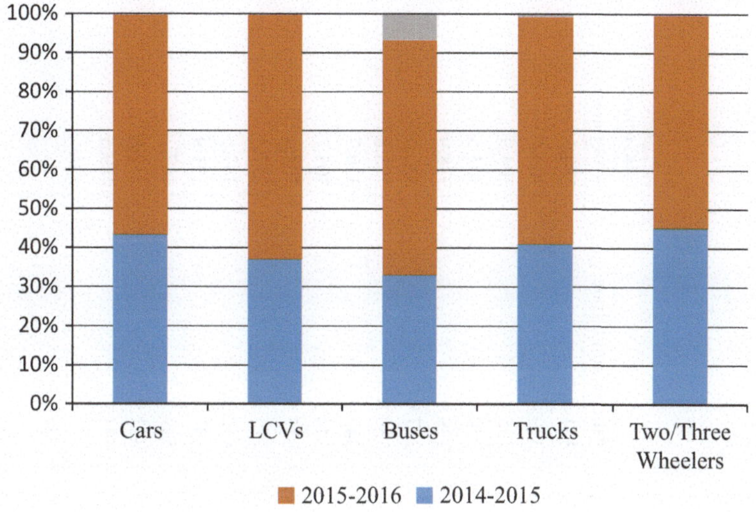

Fig. 1.11 Production of automotive industry

1.12 Other Major Products

In Tables 1.7 and 1.8, we have further provided detailed information of the manufacturing sector as primary product-wise growth from 2000 to 2016 given below; we can see the production trend of cotton yarn, cotton cloth, jute goods, vegetable ghee, cigarettes, fertilizers, cement, soda ash, caustic soda, and sugar.

In the table, there are some product trends that are declining, but overall performance is good and showing an increasing trend (Datta, Gopalakrishna-Remani, Bozan, 2015; Faisal, 2015) (Figs. 1.12 and 1.13; Table 1.8).

Three years ago, the Pakistan Federal Minister for ports and shipping, Mr. Kamran, gave Green Supply Chain Award during the conference to the more than 11 new multinational manufacturing companies working in Pakistan. That conference and awards are supported by UNEP (United Nations Environment Program) to appreciate the efforts of firms that are making green supply chain strategy (Government of Pakistan, 1988; Government of Punjab, 2002). Mainly the firms that achieve the awards include Yunus Textile Mills Limited, Future Impex, e2e

Table 1.7 Production of fertilizers

				(000 tonnes)	
Year	Super phosphate	Urea	Ammonium nitrate	Nitrophosphate	Ammonium sulfate
2000–2001	159.6	400.5.1	374.4	282.5	0
2001–2002	161.0	4259.6	329.4	305.7	0
2002–2003	147.2	4401.9	335.3	304.9	0
2003–2004	167.7	4431.6	350.4	363.5	0
2004–2005	163.1	4606.4	329.9	338.9	0
2005–2006	160.8	4806.4	327.9	356.6	0
2006–2007	148.9	4732.5	330.8	325.8	0
2007–2008	157.7	4925.0	343.7	329.7	0
2008–2009	187.4	4918.4	344.3	305.7	0
2009–2010	148.7	5056.5	345.5	304.4	0
2010–2011	173.3	4552.1	275.1	252.3	0
2011–2012	114.7	4470.1	432.3	337.6	0
2012–2013	79.3	4215.1	401.3	291.9	0
2013–2014	87.8	4930.3	519.1	447.1	0
2014–2015 (Jul–Mar)	47.3	3805.9	361.9	317.8	0
2015–2016 (Jul–Mar)	48.6	4319.9	531.1	489.5	0

Table 1.8 Production of ghee, cement, and sugar

			(000 tonnes)
Year	Ghee	Cement	Sugar
2000–2001	835	9672	2956
2001–2002	797	9935	3247
2002–2003	772	10,845	3686
2003–2004	888	12,862	4021
2004–2005	1048	16,353	3116
2005–2006	1151	18,564	2960
2006–2007	1180	22,739	3527
2007–2008	1137	26,751	4733
2008–2009	1060	28,380	3190
2009–2010	1075	31,358	3143
2010–2011	1091	28,716	4169
2011–2012	1102	29,557	4634
2012–2013	1138	31,055	5073
2013–2014	1185	31,418	5582
2014–2015 (Jul–Mar)	877	23,459	4812
2015–2016 (Jul–Mar)	931	25,900	4949

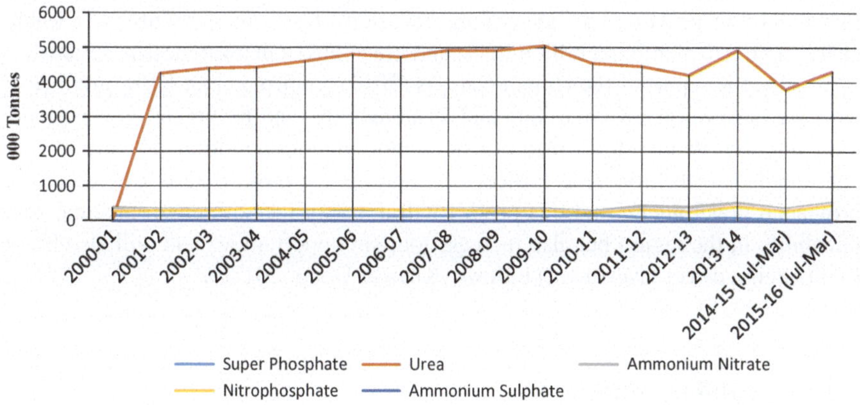

Fig. 1.12 Production of fertilizers

Supply Chain Management (Pvt) Ltd, OOCL Logistics Pakistan (Pvt) Ltd, Premier Cables (Pvt) Ltd, Pak Oasis Industries (Pvt) Ltd, Dr. Essa Laboratory and Diagnostic Centre, Anis Associates (Pvt) Ltd, Frito-Lay Pakistan, Lucky Cement Limited, and Al-Haj FAW Motors (Pvt) Ltd.

The concept of GSCM is booming in the Pakistani firms, and in the last couple of years, the federal government and provincial governments have started several “green projects” including generating electricity from the green and clean procedure and emphasizing on the green practices to protect environmental sustainability

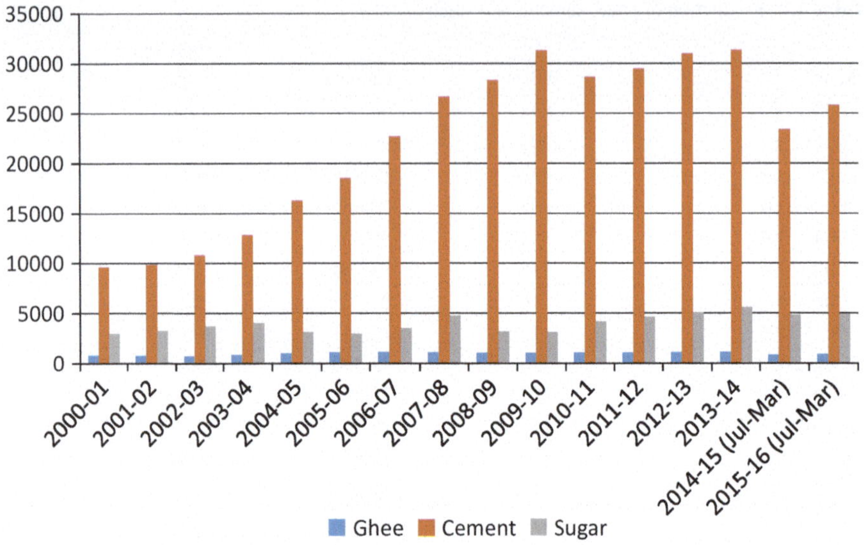

Fig. 1.13 Production of ghee, cement, and sugar

and increase the firm's profitability. The critical projects started by KPK (Khyber Pakhtunkhwa) province are generating electricity by water pressure, renewable energy sources, planting a billion trees, and green Pakistan awareness programs.

Finally, the report of the United Nations (2017) confirmed that KPK province is working hard on green environmental sustainability, and provincial governments have achieved its target of green practices and reducing pollution. The report also confirmed that "Billion Tree Project" is a successful project started by the KPK provincial government, and this project will not only increase environmental sustainability in the region but also increase tourism opportunities and will also bring FDI (foreign direct investment) inflows (Khan & Dong, 2017b).

1.13 Problem Statement

In the last decade, various studies have been conducted in many countries on an environmental friendly supply chain, green supply chain, and sustainable supply chain, and all these terms are overlapping and most of the times used interchangeably. From the perspective of the environmentally friendly supply chain, there are some drivers and motivating factors that encourage organizations to adopt and implement green and environmentally friendly practices in their supply chain (Zailani, Eltayeb, Hsu, & Tan, 2012). Observing a few of them, external and internal factors motivate firms to adopt environment-friendly practices. This research will find out the motivating factors involved in environmental friendly supply chain and also originate the effect of external and internal motivating drivers including:

- Regulation
- Social responsibility
- Predicted business advantages
- Consumer pressure

The government and regulatory authorities enforce on firms the adaptation of green practices. Many firms implemented an environmentally friendly supply chain due to the pressure of the government. On the other hand, international nonprofit organizations also create awareness to the public to avoid the usage of polluted products for healthcare and environmental sustainability (Ye et al., 2013). In developed countries, companies implement green practices in their overall business operations under the umbrella of "social responsibility." Undeniably, European countries are more conscious regarding health and environmental sustainability issues (Khan & Qianli, 2017). On the other side, in Asian countries, environmental issues become very serious as compared to European countries, and governments are pushing to enterprises the adoption of green practices in their polluted operations. The green practices implemented in business operations are not only protecting the environment but also bringing positive change in the financial performance of the firms. It is also a key reason for implementing the environmental supply chain. Many types of research have been conducted to prove that firms adopting green (Taylor, Mitra, & Datta, 2014) practices in the supply chain will significantly improve their financial performance.

Consumer pressure playing a significant role in encouraging firms to adopt green practices in the end-to-end supply chain. Firms manufactured products for consumers, and if consumers are not willing to buy the products, so in this situation, firms cannot survive in the long term. In today's world, consumers are very aware of every information; the Internet also plays a critical role, and consumers can get all information in a click (Kim & Min, 2011). Finally, consumer awareness and governmental laws create pressure on firms to adopt GSCM (green supply chain management). Environmentally friendly supply chains are a new evolving concept, and the adoption of external and internal sustainable factors and driver organizations can minimize the destructive effect on the environment, and even they can restrict the negative effect on the environment (Giovanni, 2012). There are also some barriers and hurdles, which can create difficulties and problems for the companies during the implementation phase of green practices, including:

- Heavy investment
- Employee training
- Lack of top management involvement

In the implementation of GSCM, firms need financial support to bear substantial investment. Many firms have a willingness to adopt green supply chain management in their business activities, but financial scarcity becomes the main barrier. Secondly, training of employees also required huge investment and time. Because "change never comes dramatically," employees also need time to learn new processes of green supply chain and their implementation in the manufacturing processes. Finally, due to the lack of top management support or involvement, as per several researchers, many green projects become drain (Khan & Qianli, 2017) over firms' profit. The top management seriousness and commitment toward green concept is an initial step toward the journey of green supply chain management (Wei et al., 2012; Khan & Dong, 2017a).

This research will emphasize on barriers, motivating factors, or drivers involved in the implementation of environmental friendly supply chain management and also discover the latest green supply chain management practices, motivating drivers, and barriers faced by the organizations in the execution of green supply chain management.

1.14 Research Objectives

- The research purpose is to find out motivating factors and drivers of green supply chain management, which motivate the organizations to adopt environmentally friendly practices in their business operations.
- This study will also help to find out barriers and hurdles faced by the firms during the execution of green supply chain management practices.
- This study also will provide optimum possible solutions to all these barriers and hurdles faced during the implementation phase of green supply chain practices.

1.15 Research Scope

The scope of research means the gap in the literature review from the perspective of the research conducted area. Organizations can minimize or reduce the harmful effects of supply chain operations over the environment by adopting sustainable and environmentally friendly supply chain practices. The researcher gives the idea that organizations face problems and hurdles when they switch their conventional supply chain to an environmentally friendly and green supply chain. This research will not only help the organizations regarding the smooth implementation of environment-friendly practices in their supply chain operations but will also help to find out driving forces and motivating factors of green supply chain and barriers/hurdles faced by companies during the execution phase of environmentally friendly practices in their supply chain.

1.16 Research Method

The researcher adopts PLS-SEM for the quantitative part of this study. The PLS-SEM is a collection of statistical techniques that allow a set of relationships between one or more exogenous variables, be it continuous or discrete, to be examined.

PLS-SEM is a causal modeling approach, maximizing the explained variance of the dependent latent constructs. It is also efficient for theory building, theory testing, and exploratory analysis. Furthermore, the following are the key benefits of PLS-SEM:

- Useful for preliminary theory building and causal-predictive analysis.
- PLS-SEM is a suitable method and works better in both cases (small sample size and big sample size).
- PLS-SEM is a causal modeling method that also enhances the validity and reliability of the results.
- It does not have the capability for comparison of the observed and proposed covariance matrices.
- Reliable for exploratory analysis, which is required for theory building.
- It can be used for both confirmatory and exploratory analyses.
- Avoids indeterminacy by composing constructs from factor scores and using these in subsequent calculations, yielding explicit factor scores.
- It can be used in models with formative (or a mix of reflective and formative) indicators.
- It is robust to violations of multivariate normal distributions.

The main advantage of the PLS-SEM is that it makes fewer assumptions about how the data are distributed and is more robust to smaller sample size limitations and offer in-depth analysis. PLS path modeling also flexibly allows specifying both formative and reflective constructs (latent variables) in one model. However, the two (CB-SEM and PLS-SEM) are different and are useful for different research purposes and questions. Think of PLS path modeling as good for prediction (which factor predicts what?) and/or hypothesis generation (Hair, 2014). On the other hand, the researcher has adopted "unstructural interviews" for the qualitative part of this research. The researcher used a purposive sampling method to identify the target respondents. Besides, "unstructural interviews" provide more in-detailed knowledge and also are recommended for analyzing phenomena. This research will identify barriers based on "unstructural interviews," while drivers of GSCM will be identified through the qualitative statistical analysis.

1.17 Thesis Structure

This thesis report mainly covers seven chapters explained as follows:

In the first chapter, we have discussed the background of environmental friendly supply chain management as a motivation to the thesis, problem statement, or issues of green supply chain management needed to be resolved, objectives of the research, scope of the research, and research method.

The second chapter comprises the history of Pakistani manufacturing industry, dynamics of manufacturing industry, and the trend of growth since the last decade. In this research, we will discuss the barriers, problems, and motivating factors involved in the environmentally friendly supply chain from the perspective of Pakistani manufacturing industries.

The third chapter will cover the literature reviews of the green supply chain, environmentally friendly supply chain from the perspective of manufacturing industries. Besides, previous empirical study results and methodologies, case studies results, and approaches are used to solve the problems. Further, we will also develop and discuss hypothesis building in detail.

The fourth chapter discusses the methodologies and approaches used to achieve the objective and goal of this research. The chapter will also cover:

- Research approach: This research approach is going to be conducted in mix approach, where we use statistical analysis techniques and tools; we will also describe the situation and a phenomenon with in-depth interviews with supply chain managers.
- Research purpose: This research will cover exploratory and explanatory purposes. As far as green supply chain concerns, researcher wants to explore and explain the motivating drivers and barriers which organizations face in the implementation of environmentally friendly practices.
- Research design: In this research, we are going to use both phenomenological and correlational design.
- Data collection instrument: In this research, for data collection, the researcher has selected the questionnaire method and in-depth interviews from senior management of the supply chain.
- Sampling techniques: The researcher is going to use purposive sampling for this research to achieve the critical objective of the research.
- Data source: The researcher is going to use primary data for this research. Moreover, data will be collected from supply chain professionals and managers.
- Data instrument: The researcher is going to use a questionnaire (Likert scale), and in-depth interviews will also be conducted with supply chain directors and managers.
- Research model and model hypothesis.

The fifth chapter discusses the statistical tests and analyses, including descriptive statistics, correlation matrix, estimation results, and structural model results. Besides, this chapter will discuss the results of each hypothesis in detailed and explain the impact of the results over the research model.

The sixth chapter comprises the in-depth discussion on model results and detailed discussion on unstructured interview results. Finally, the research results have been supported by other researchers' findings.

Furthermore, chapter seven covers the conclusion of this research and innovation points and provides recommendations from the perspective of the Pakistani manufacturing industry.

The outline of the thesis has been illustrated in Table 1.9.

Table 1.9 Thesis outline

Problem statement and literature support			Research methodology	Data analysis results			Discussion and conclusion	
Discussed in Chap. 1	Discussed in Chap. 2	Discussed in Chap. 3	Discussed in Chap. 4	Discussed in Chap. 5			Discussed in Chap. 6	Discussed in Chap. 7
Background of GSCM in Pakistan Environmental laws in Pakistan Problem statement Object of the research Scope of the research	History of Pakistan manufacturing industry Dynamics of manufacturing industry Role of manufacturing industry in economic growth Development and growth of manufacturing firms in Pakistan	Literature review on green supply chain management Latest techniques of green supply chain management in the perspective of manufacturing industry Development of research hypothesis	Research approach Research design Data collection instrument Sampling techniques Data source Research modeling with hypothesis	Survey responses analysis Demographic profile of respondents Data screening and initial analysis Normality of data Common method bias Reliability and validity analysis Descriptive statistics and matrix correlation KMO and Bartlett's test	CFA analysis Measurement model for testing Outer loading of indicators Outer weight values Inner VIF values Convergent validity of constructs Discriminant validity by Fornell-Larcker criterion Cross-loading analysis Heterotrait-monotrait ratio (HTMT) analysis	R-square and adjusted R-square The f-square values The result of structural model Result of hypothesis testing Model fit statistics Additional model fit statistics Predictive relevance (Q2) Drivers and motivating factors of GSCM	In-depth discussion on model result In-detailed discussion on unstructured interview results Finally, the research results have been supported	Implication of research and conclusion Key findings of research Innovation points and recommendations Research limitations and future research

References

Brandenburg, M. (2015). Low carbon supply chain configuration for a new product – A goal programming approach. *International Journal of Production Research, 7543*(April), 1–23. https://doi.org/10.1080/00207543.2015.1005761

Chari, N., Diallo, C., & Venkatadri, U. (2014). State of the art on performability across the sustainable value chain. *International Journal of Performability Engineering, 10*(6), 543–556. Retrieved from https://www.scopus.com/inward/record.uri?eid=2-s2.0-84906859309&partnerID=40&md5=beda4b9ade6edaaf1e75e7923cacfc1d

Datta, P., Gopalakrishna-Remani, V., & Bozan, K. (2015). The impact of sustainable governance and practices on business performance : An empirical investigation of global firms. *International Journal of Sustainable Society, 7*(2), 97–120.

Faisal, (2015). Mind meets machine, Science Journal, 350(6262), 746, https://doi.org/10.1126/science.aad0787, https://science.sciencemag.org/content/350/6262/746.1

Giovanni, P. D. (2012). Do internal and external environmental management contribute to the triple bottom line?, *International Journal of Operations and Production Management, 32*(3), 265–290.

Global Industry Analysts, (2013) Influencer Driven AI Powered Real-Time Market Intelligence, https://www.strategyr.com/ [Accessed on March 2019].

Government of Pakistan, (1988). Mission and goals to be achieved, http://www.pakistan.gov.pk/ [Accessed on July 2019].

Government of Punjab, (2002). Punjab Portar All About Punjab At One Place, https://www.punjab.gov.pk/ [Accessed on September 2019].

Gruner, R. L., & Power, D. (2017). Mimicking natural ecosystems to develop sustainable supply chains: A theory of socio-ecological intergradation. *Journal of Cleaner Production, 149*, 251–264. https://doi.org/10.1016/j.jclepro.2017.02.109

Habidin, N. F., Fadzlin, A., Zubir, M., & Fuzi, N. M. (2017). Critical success factors of sustainable manufacturing practices in the Malaysian automotive industry. *International Journal of Sustainable Engineering, 11*, 217–222. https://doi.org/10.1080/19397038.2017.1293185

Hair, (2014) Multivariate Data Analysis, Seventh Edition, ISBN 10: 1-292-02190-X, British Library Cataloguing-in-Publication data, British Library, Pearson, USA.

Jakowczyk, M., Quariguasi, J., Neto, F., Gibson, A., & Van, L. N. (2017). Understanding the market for remanufactured products : What can we learn from online trading and Web search sites ? *International Journal of Production Research, 7543*(January). https://doi.org/10.1080/00207543.2016.1275872

Keehung Lai & Wong, (2012). Green logistics management and performance: some empirical evidence from Chinese manufacturing exporters. *Omega, 40*(3), 267–282.

Khan, S. A. R. (2019). The nexus between carbon emissions, poverty, economic growth, and logistics operations-empirical evidence from southeast Asian countries. *Environmental Science and Pollution Research, 26*, 13210–13220. https://doi.org/10.1007/s11356-019-04829-4

Khan, S. A. R., & Dong, Q. (2017a). Impact of green supply chain management practices on firms' performance: An empirical study from the perspective of Pakistan. *Environmental Science and Pollution Research, 24*(20), 16829–16844.

Khan, S. A. R., & Dong, Q. (2017b). Does national scale economic and environmental indicators spur logistics performance? Evidence from UK. *Environmental Science and Pollution Research, 24*(34), 26692–26705.

Khan, S. A. R., Dong, Q., & Zhang, Y. (2018). The impact of green supply chain practices in business performance: Evidence from Pakistani FMCG firms. *Journal of Advanced Manufacturing Systems, 17*(2), 267–275.

Khan, S. A. R., Golpira, H., Zhang, Y. (2018). The Importance of Advanced Information Technology and Green Vehicles in Supply Chain Management, International Conference on Computer, Communications and Mechatronics Engineering, ISBN 978-60595-611-4.

Khan, S. A. R., Dong, Q., Zhang, Y., & Khan, S. H. (2017). Research on Decision-making of Green Reverse Logistics in Enterprises: A Case Study on Electronic Products Manufacturers from the Perspective of South Africa, AMEIT 2017, ISBN 978-1-60595-457-8.

Khan, S. A. R., Dong, Q., Zhang, Y., & Khan, S. S. (2017). The impact of green supply chain on enterprise performance: In the perspective of China. *Journal of Advanced Manufacturing Systems, 16*(3), 263–273.

Khan, S. A. R., & Qianli, D. (2017). Impact of green supply chain management practices on firms' performance: An empirical study from the perspective of Pakistan. *Environmental Science and Pollution Research, 24*, 16829–16844. https://doi.org/10.1007/s11356-017-9172-5

Khan, S. A. R., Qianli, D., SongBo, W., Zaman, K., & Zhang, Y. (2017). Environmental logistics performance indicators affecting per capita income and sectoral growth: Evidence from a panel of selected global ranked logistics countries. *Environmental Science and Pollution Research, 24*(2), 1518–1531. https://doi.org/10.1007/s11356-016-7916-2

Khan, S. A. R., Zaman, K., & Zhang. (2016). The relationship between energy-resource depletion, climate change, health resources, and the environmental Kuznets curve: Evidence from the panel of selected developed countries. *Renewable and Sustainable Energy Reviews, 62*(9), 468–477.

Khan, S.A.R., Zhang, Y. (2019) Strategic supply chain management, EAI/Springer innovations in communication and computing series, Springer Publisher, ISBN 978-3-030-15058-7, https://www.springer.com/gp/book/9783030150570

Khan, S. A. R., Zhang, Y., Anees, M., Golpîra, H., Lahmar, A., & Dong, Q. (2018). Green supply chain management, economic growth, and environment: A GMM based evidence. *Journal of Cleaner Production, 185*(6), 588–599.

Kim, I., & Min, H. (2011). Measuring supply chain efficiency from a green perspective. *Management Research Review, 34*(11), 1169–1189. https://doi.org/10.1108/01409171111178738

Lai, K., & Wong, C. W. Y. (2012). Green logistics management and performance: some empirical evidence from Chinese manufacturing exporters. *Omega 40*(3), 267–282.

Lee, K. H., & Wu, Y. (2014). Integrating sustainability performance measurement into logistics and supply networks: A multi-methodological approach. *British Accounting Review, 46*(4), 361–378. https://doi.org/10.1016/j.bar.2014.10.005

Liu, L., Tang, M., & Xue, F. (2012). The impact of manufacturing firms ' green supply chain management on competitive advantage. *Applied Mechanics and Materials, 475*, 3349–3354. https://doi.org/10.4028/www.scientific.net/AMR.472-475.3349

Lorenzoni & Lipparini (1999). The leveraging of interfirm relationships as a distinctive organizational capability: a longitudinal study. *Strategic Management Journal, 20*(4), 317–338.

Meneghetti, A., & Monti, L. (2014). Greening the food supply chain: An optimization model for the sustainable design of refrigerated automated warehouses. *International Journal of Production Research, 7543*(June 2015), 1–21. https://doi.org/10.1080/00207543.2014.985449

Meneses, Y. E., Stratton, J., & Flores, R. A. (2017). Water reconditioning and reuse in the food processing industry: Current situation and challenges. *Trends in Food Science and Technology, 61*, 72–79. https://doi.org/10.1016/j.tifs.2016.12.008

Monczka, R. M., Handfield, R. B., Giunipero, L. C., & Patterson, J. L. (2009). Purchasing and supply chain management. Cengage Learning.

Nakajima, K., Nansai, K., Matsubae, K., Tomita, M., Takayanagi, W., & Nagasaka, T. (2017). Science of the total environment global land-use change hidden behind nickel consumption. *Science of the Total Environment, 586*, 730–737. https://doi.org/10.1016/j.scitotenv.2017.02.049

Pakistan Bureau of Statistics. (2016). Pakistan Statistical Year Book 2016, http://www.pbs.gov.pk/content/pakistan-statistical-year-book-2016 [Accessed on July 23, 2019].

Savita, K. S., Dominic, P. D. D., & Ramayah, T. (2014). The adoption of green information technologies and systems as a driver within green SCM, International Conference on Computer and Information Sciences (ICCOINS), Kuala Lumpur, pp. 1–6.

Shin, Y., & Thai, V. V. (2016). A study of the influence of sustainable management activities on customer satisfaction and long-term orientation in the shipping industry : Evidence from users of Korean flagged shipping service. *International Journal of Shipping and Transport Logistics, 8*(1), 1–20.

State Bank of Pakistan. (2016). Financial Stability Review-2016, http://www.sbp.org.pk/FSR/2016/index.htm, [Accessed on March 11, 2019].

Steinhilper, R. (2001). Recent trends and benefits of remanufacturing: from closed loop businesses to synergetic networks, Proceedings Second International Symposium on Environmentally Conscious Design and Inverse Manufacturing, Tokyo, Japan, 2001, pp. 481–488.

Taylor, P., Marshall, D., Mccarthy, L., Heavey, C., & Mcgrath, P. (2014). Production planning & control : The management of operations environmental and social supply chain management sustainability practices : Construct development and measurement, (November), 37–41. https://doi.org/10.1080/09537287.2014.963726

Taylor, P., Mitra, S., & Datta, P. P. (2014). Adoption of green supply chain management practices and their impact on performance : An exploratory study of Indian manufacturing firms, (May), 37–41. https://doi.org/10.1080/00207543.2013.849014

Tibben-Lembke, R. S., & Rogers, D. (2002). Differences between forward and reverse logistics in a retail environment. Supply Chain Management, 7(5), 271–282. https://doi.org/10.1108/13598540210447719.

The British Standards Institution, (2010). Annual Report and Financial Statement 2010, https://www.bsigroup.com/Documents/about-bsi/financial-performance/2010/bsi-financial-performance-2010.pdf.

Thierry, M., Salomon, M., Nunen, J. V., Wassenhove, L. V. (1995). Strategic Issues in Product Recovery Management, California Management Review, 37(2), 114–134.

United Nations. (2016). Global Sustainable Development Report, 2016 Edition, https://sustainabledevelopment.un.org/content/documents/2328Global%20Sustainable%20development%20report%202016%20(final).pdf, [Accessed 2017 03, July].

United Nations. (2017). Sustainable Development Goals Report, https://www.un.org/development/desa/publications/sdg-report-2017.html [Accessed on March 2019].

Van, A. J., Kuss, D. J., Griffiths, M. D., Shorter, G. W., Schoenmakers, M. T., Mheen, D. V. D. (2014). The (co-)occurence of problematic video gaming, substance use, and psychosocial problems in adolescents, *Journal of Behavioral additions, 3*(3), 157–165.

World Bank, (2015). World Bank Annual Reports 2015, https://www.worldbank.org/en/about/annual-report-2015 [Accessed on March 18, 2018].

Ye, F., Zhao, X., Prahinski, C., & Li, Y. (2013). The impact of institutional pressures, top managers ' posture, and reverse logistics on performance — Evidence from China. *International Journal of Production Economics, 143*(1), 132–143. https://doi.org/10.1016/j.ijpe.2012.12.021

Yusop, N. M., Wahab, D. A., & Saibani, N. (2016). Realizing the automotive remanufacturing roadmap in Malaysia: Challenges and the way forward. *Journal of Cleaner Production, 112*, 1910–1919. https://doi.org/10.1016/j.jclepro.2015.03.072

Zailani, S. H. M., Eltayeb, T. K., Hsu, C. C., & Tan, K. C. (2012). The impact of external institutional drivers and internal strategy on environmental performance. *International Journal of Operations & Production Management, 32*(6), 721–745.

Chapter 2
Empirical Studies on Green Supply Chain Management

2.1 Introduction

The main crack of the idea of green supply chain management is to protect environmental sustainability, but as per the researchers, firms also adopt GSCM to increase their financial and operational performance and build a competitive edge in the domestic and international arena (Halldórsson & Kovács, 2010). Undeniably, GSCM is valuable for firms' environmental and economic performance, while the benefits of GSCM cannot be measured in the short term. Environmental friendly supply chain and sustainable supply chain benefits can be measured in the long run on countries' economic growth and foreign direct investment inflows (Zaman & Shamsuddin, 2017).

2.2 Development in GSCM

Lee et al. (2013) investigate the relationship between supply chain performance and pressure affecting environmental SC. The author collected the data of the green supply chain through the questionnaire as well as for the test structural equation model, and factor analysis of confirmatory has been used. The research findings show that executions of green practices in the supply chain sustain or increase performance and competitive edge. Increase the flexibility of supply chain management reduces the resources through the improvement of output and decrease cost factors. Green supply chain management enterprises have policies inclined to increase the flexibility of the supply chain and also enhance profits.

Laosirihong thong et al. (2013) investigate the execution of environmental supply chain development of reactive and proactive practices and analyze green supply chain impact over economic, environmental, and intangible performance. The

S. A. R. Khan, *The Critical Success Factors of Green Supply Chain Management in Emerging Economies*, EAI/Springer Innovations in Communication and Computing,
https://doi.org/10.1007/978-3-030-42742-9_2

limitation of the study is the cultural interaction of firms between suppliers and key customers. The sample data was from Thailand ISO 14001 certified 190 s organization and apply in test multivariate linear regression. Research results show that proactive practices in reverse logistics do not have a significant impact on the environmental supply chain, as well as had adoption of a low level. Moreover, reactive practices in reverse logistics due to (threat of regulation and legislation) had a significant result as companies' intangible performance, economic, and environmental Green et al. (2012) Examining the impact of green supply chain production on performance. The researchers developed a model of GSCM practices and created a link between the supply chain partners (including suppliers and customers) and manufacturer support. Supply chain sustainability. Research results show that economic and environmental performance improved by the adoption of green manufacturing practices. Also the researcher did collect data from 159 manufacturing companies' managers and used the methodology of the structural equation model. The limitation of studies is an exploratory, empirical investigation of the firms' performance from environmental practices.

2.3 Development of Hypothesis

Mainly the green supply chain incorporates the critical idea of social, environmental, and economic performance (Khan & Qianli, 2017). The literature connected to the green supply chain is comparatively well-organized and developed. The key focus of our research is the environmental and economic performance factor of sustainability. Undeniably, the focus of environmental sustainability has been shifted from the firm level to the SC (supply chain) level. In the light of Seuring (2004), green supply chain is the managerial integration of information and material flows by the SC to fulfill the requirement of customers for environmental-friendly products manufactured by green materials, processes, design, and packaging, which reduce environmental degradation. Several types of research have been done on the topic of the green supply chain, sustainable supply chain, and environmental-friendly supply chain, but the results are still inconclusive of all previously published research papers. This is the fundamental motivation and gap to fill up through this study, which we conduct on the Pakistani manufacturing firms. In the following subheadings, we will develop a one-by-one hypothesis with supporting published literature review.

2.3.1 Internal Environmental Management and Green Information Systems

IEM is a prerequisite for the adoption of green practices in the end-to-end supply chain management. The role of IEM is significantly essential in improving firms' performance in terms of senior management/top leadership commitment and cross-functional collaboration; the commitment of senior management is the primary factor, even if firms are at a different level of green project execution (Hsu & Hu, 2008).

Whenever companies implement green supply chain or environmental sustainability as imperative of strategic, then the company may process to build GIS (green information system) capabilities. No doubt, IS (information systems) is a crucial contributor to the maintenance and long-term survival of SCs (supply chains). The implementation of green practices is an imperative strategic need for companies to build and execute end-to-end and robust GIS (green information systems). A long-term strategic emphasize on green supply chain management requires the essential to monitor activities and operations performed by the manufacturing selling and purchasing process to ensure a reduction in environmental degradation and increase or enhanced environmental sustainability. IEM is a systematic process containing of a complete set of green supply chain/environmental sustainability policies, valuation of internal policies and their impacts on environmental sustainability, plans of action, quantifiable environmental sustainability targets, and measure through regular auditing of these factors (Jabbour, Frascareli, & Jabbour, 2015).

Internal environmental management practices create positive effects on environmental performance and are a key contributor to the reduction of environmental degradation and carbon emission to enhanced manufacturing firms' performance and environmental sustainability. The IEM is written environmental policies and objective, namely, senior and middle management commitment towards green projects, sophisticated (IT) information systems to monitor environmental regulations and laws, supplier environmental plan, firm certification of ISO 14001 and 9000, strict regulatory compliance audits, and assignment of management representatives (Sharma, Chandna, & Bhardwaj, 2017). On the other hand, internal environmental management is the leading performance factor which mainly based on senior management commitments with green projects.

The systems of information sharing and monitoring are not just allowed and enablers of interconnectedness; they (Welty and Becerra-Fernandez, 2001) may also be used to improve commitment, collaboration, and trust among the end-to-end supply chain members. ISs (information systems) are used by firms for offering techniques, tools of collaboration, and trust building. Undeniably, the support of top leadership or senior management is an essential precursor to fully achieve the execution of information systems. If firms are willing to adopt green practices in their supply chain, so information systems should incorporate and (Green et al., 2012) integrate environmental efforts because of the IEM. Accordingly, we build a hypothesis that IEM (internal environmental management) enhanced GIS (green information systems).

H-1: Internal environmental management significantly and positively improved green information systems.

2.3.2 Internal Environmental Management and Green Manufacturing

Green manufacturing practice cannot be implemented successfully in the firm without the senior management commitment, while IEM is a backbone of the green supply chain management. Senior management influences on supply chain partners to adopt green practices in their business practices, and green manufacturing is based on the master production plan (MPP), which is approved by the firm's senior leadership (Zhu, Sarkis, & Lai, 2007). Internal environmental management practices implemented based on the senior leadership support on inter-functional collaboration and cooperation to achieve in-house environmental sustainability goals such as reduction in hazardous chemicals, toxic chemicals, and materials, reduction in water and electricity consumption on ISO 14001 certification, on the fulfillment of environmental, on internal audit performance and the legal aspects (Jabbour, Jabbour, Govindan, Kannan, & Arantes, 2014).

Undeniably, IEM is one of the significant green supply chain management practices that firms must implement to enhance their manufacturing processes and environmental performance, and IEM is chiefly enhances firm's performance through top management commitment and cross-functional cooperation within the firm and with supply chain partners (Wei, Ju, & Angkasa, 2016). Zhu and Sarkis (2004) described that the quality of products has a strong and positive association with environmental management performance. If a company has an active environmental strategy, it will allow and support firms in environmental design, green logistics and distribution, collaboration with customers, suppliers involvement in product development, and building green manufacturing strategies (Khan & Qianli, 2017). Explain to green manufacturing involves planning and control, reducing the consumption of energy and material exploitation, reduction of waste from the end-to-end processes of manufacturing. GSCM backing firms in order to enhance environmental and economic performance and to offer staff conducted environmental-based training. If a firm is serious about adopting green supply chain practices, the company must need to adopt internal environmental management to support the complete process of green supply chain; otherwise, it will be only garbage-in and garbage-out (Zhu & Sarkis, 2004). Green manufacturing practices and greening processes of manufacturing cannot be achieved without the senior and middle management support.

Mainly environmental-friendly manufacturing involves the implementation of the best resources, which in the long term may achieve competitive advantage through high-level quality products. In the same way, firms reduce their costs and remove waste from the end-to-end supply chain system by using technologies of green processes and green manufacturing. On the other hand, lean manufacturing and environmental-friendly manufacturing both concept objective is the removing of waste, defects and enhance the efficiency of manufacturing, while green manufacturing is more talk about environmental sustainability, which also achieves more superior environmental and financial performance (Khan & Qianli, 2017).

On the other hand, senior management is mainly responsible for increasing shareholder profit and, through strategic position and long-term strategies, determining the direction of the firm, including vision, and defining its values and strategic intent. Generally, the senior management commitment towards green supply chain practices is due to commercial, technological, and financial reasons (Bowen, 2001). Therefore, we develop the following hypothesis to test the relationship between IEM (internal environmental management) and GIS (green information systems).

H-2: IEM (internal environmental management) has a significant and positive association with GM (green manufacturing).

2.3.3 Internal Environmental Management and Green Distribution

Internal environmental management has a direct and significant effect on the implementation of green practices in distribution, which is a core part of end-to-end supply chain management. Usually, in the internal environmental management, the following variables are accounted for the commitment of the senior leadership towards green practices in distribution, legal and environmental requirements, and ISO 14001 (Jabbour et al., 2014). The principal objective of the firms to implementation of green practices in distribution is to increase environmental sustainability and reduce systems cost. As for the adoption of green practices in the supply chain, it means that the firm needs to pay close attention to environmental factors, at the same time mind (Hajmohammad, Vachon, Klassen, & Gavronski, 2013) upstream and downstream collaboration.

Mainly the green practices included green distribution, green manufacturing, green design, green purchasing, and green marketing (Bakhodirov & Peng, 2009). The manufacturing companies may minimize the harmful environmental effect on environmental sustainability through the adoption of green practices and by increasing the level of investment in cleaner technologies, shifting investment from pollution control towards the prevention of pollution (Khan and Dong, 2017a, b; Khan and Zhang, 2019). Undeniably, many companies primarily emphasize on their distribution networks and manufacturing processes. Conversely, despite its broad recognition, companies have only recently started to involve in green purchasing that spans the supply chain (upstream) and has identified that a concerted effort is essential to face the challenges of environmental sustainability (Khan, Dong, Zhang, Khan, 2017). The impact of green distribution and cooperation between marketing and logistics had found a positive effect on performance (Khan et al., 2017a, b). In the same way, in the container shipping industry, researchers argue that internal green supply chain practices and integration are significantly and positively correlated with company competitiveness and green performance.

Recently several companies have understood the importance of green distribution/warehousing to reduce energy consumption and cost. However, the starting cost and investment of time to convert to this kind of distribution/warehousing scare many companies (Khan, Dong, & Zhang, 2016, 2017, 2018). To reduce the cost of distribution systems and enhance environmental sustainability, several warehousing and distribution firms are adopting green practices in their supply chain to reduce the harmful effect of carbon dioxide and greenhouse gases (Rostamzadeh, Govindan, Esmaeili, & Sabaghi, 2015). As environmental awareness is increasing, firms are adopting green practices in their business to maintain their market share, customer satisfaction, and profitability. The green practices range from reducing the amount of GHG (greenhouse gases), carbon emissions (carbon dioxide), and fossil fuel consumption. The green distribution systems refer to warehousing which include storage, order picking, processing, packaging, delivery to the customer, and tracking forward and backward chain (Mohammed, Selim, Hassan, & Syed, 2017). Accordingly, we build a hypothesis that IEM (internal environmental management) has a significant and positive effect on GD (green distribution).

H-3: IEM (internal environmental management) has a significant and positive effect on GD (green distribution).

2.3.4 Internal Environmental Management and Green Transportation

In the green supply chain projects, firms cannot ignore to the adoption of green practices in the transportation, undeniably green transportation plays a significant role, and green practices are unable to deploy in transportation without the top leadership commitment towards green projects (Dongmin, Dingyi, Youfang, & Bin, 2016). Transportation emits massive carbon dioxide and greenhouse gases, which create a significant and negative effect on environmental sustainability in the long term. Rostamzadeh et al. (2015) confirmed that transportation emits 23% of carbon and 15% of GHG (greenhouse gases), which are the cause of climate change and global rise in temperature.

The IEM (internal environmental management) represents the senior leadership commitment towards green supply chain management projects. The senior management cannot ignore the transportation movement and negative effect on the environment if they have decided to adopt green practices in the end-to-end supply chain management. Generally, internal environmental management emphasizes the adoption of green practices, namely, green manufacturing, green design, green distribution, and green transportation (Rostamzadeh et al., 2015). Green transportation has a significant and positive correlation with a competitive edge and environmental performance, which are the bigger goal of the firm in the implementation of green practices. Usually, in the green transportation, firms use eco-efficient fleet such as energy-efficient vessels and green fuels like low sulfur content, ensure proper maintenance of a vehicle, provide training to the employees, and achieve green transportation certification from government/regulatory authority which are the key points needed to be implemented in green transportation. On the other hand, green transportation not only increases environmental sustainability but also enhances and develops a competitive edge, firm image, and reputation, and most times (Yang, Lu, Jing, & Bernard, 2013) the government provides subsidies and tax exemption to firms.

Environmental-friendly transport refers to the adoption of green practices in transportation to reduce the harmful effect of transportation operation on environmental sustainability and to mitigate climate change. Khan et al. (2020)

argues that green transportation evaluation can be based on many components which are the reduction of fossil fuel burning/consumption, employees' training, vehicle maintenance, and usage of pollution-free vehicle. Generally, transportation sustainability is broadly being measured by transportation system efficiency and effectiveness as well as climate and environmental impacts of the system.

Transportation has appeared as a critical factor behind the emission of harmful pollutants, including carbon emissions, nitrogen oxide, and particulate matter (PM2.5). Greenhouse gas emissions increased due to the burning of non-green fuels like fossil fuels, which create serious climate issues and environmental sustainability problems that post health risks and significantly harmful for the environment and human lives (Demir, Bekta, & Laporte, 2014). The aim of green transportation is to reduce fossil fuel consumption and/or fuel burning to increase environmental sustainability and enhance or build a firm's competitive edge in the international arena. Therefore, we initiated the following hypothesis to measure the correlation between IEM (internal environment) and GT (green transportation).

H-4: IEM (internal environmental management) has a significant and positive correlation with GT (green transportation).

2.3.5 *Internal Environmental Management and Ecological Design*

The internal environmental management has a significant and positive effect on the ecological design and emphasizes the firm's policies to adopt green practices in their supply chain management. Mainly IEM (internal environmental management) is based on commitment of senior leadership towards execution of green practices to attain green supply chain management goal, which is increased to the environmental sustainability and reduce harmful effect on the environment, senior leadership making green practices as their firm policy since mission statement and policy of quality are decided by the senior management (Sharma et al., 2017). IEM also covers supplier environmental plan, ISO certification of 9000 and 14,001. Zhu et al. (2007) conducted the research green supply chain management practices in the context of Chinese firms. The results show that ecological design has a significant and vital role in the successful adoption of green supply chain management, while ecological design enhanced firm economic and environmental performance with improved image and reputation in the international arena.

During the implementation of green supply chain management, senior/top management have a crucial role in keeping all the operations smooth, and middle management start to find green supplier on the commitment of senior leadership; green suppliers not only provide green materials and components but also share their expertise with manufacturing, and most times manufacturing firms build strong relationship with green suppliers for building competitive edge in the market (Zhu et al., 2007). Environmental management programs to avoid a situation where vendors will delay or stop manufacturing due to environmental problems, that is, to sustain business continuity and minimized risk (Gualandris & Kalchschmidt, 2015).

Firms concerned with materials and components choice have started implementing a comprehensive environmental-friendly design approach: minimizing environmental impacts of manufacturing, recycling materials after the PLC (product life cycle), reduction in energy, and fossil fuel consumption. The ecological design has been defined as a complete process that handles environmental attributes as design objectives, product development to take account of environmental sustainability. Undeniably, environmental issues include a broader set of stakeholders and (Polonsky and Ottman 1998) provide training to the employees and build active collaboration and coordination among green suppliers (upstream) and green distributor (downstream). Ecological design refers to reduction of energy, hazardous materials, toxic chemicals design product, and green packaging/labeling, which can be recyclable, reusable, and easy to re-manufacture, while the key objective of ecological design practices cannot be successfully achieved without the involvement of the supplier during the green product development. Accordingly, we build a hypothesis that IEM (internal environmental management) has a significant and positive effect on product ED development (ecological design).

H-5: IEM (internal environmental management) has a significant and positive effect on product ED (ecological design).

2.3.6 Green Information System and Green Manufacturing

The successful adoption of the green supply chain management practices mainly depends on the capability of the firm's information systems to capture and monitor data related to the environmental performance and sustainability efforts and output of the firm's manufacturing processes. The supply chain practices cannot be implemented without the proper implementation of GIS (green information system) in the firm, and the information sharing and collaboration with upstream and downstream become stronger due to GIS. The green system represents the backbone of green management efforts by backing the company's IEM and by meeting the reporting requirement of the many supply chain members, while the senior management can enforce on the firm to adopt green information system, and this system brings positive change in green manufacturing to reduce their harmful effect on environmental sustainability (El-Gayar and Fritz, 2006). On the other hand, the traditional supply chain emits heavy air and water pollution, which also increased environmental risk in order of heavy pollution, climate change, and global warming issues.

Green information sharing provides more meaningful information sharing with suppliers and customers in terms of green manufacturing processes and practices; green information system brings information which is necessary to make suitable and appropriate decisions in the manufacturing processes. Green manufacturing talks about the reduction of waste, energy consumption, and fossil fuel consumption (Martí, Tancrez, & Seifert, 2015) and utilizes the sources of renewable energy and increased the firm's environmental and economic performance. The initial critical phase in the embarking upon the manufacturing of green products is selecting green suppliers. During the selection of green suppliers, it is necessary to allow effective technique of modeling capable of offering information on how supplier behavior with environmental factors and economic factors (Orji & Wei, 2015).

Green manufacturing followed all the processes, which are reducing harmful effects on environmental sustainability. Usually, they are considered as manufacturing techniques and approaches that efficiently use components, raw materials, and energy with less environmental impact while manufacturing moderate to no pollution and waste (Scur & Barbosa, 2017). The green manufacturing processes are more efficient to adopt and allow product to re-manufacture or recycle. The essential advantage of the green supply chain practices are to increase environmental sustainability with more significant economic benefits, and these can be achieved (Scur & Barbosa, 2017) through the adoption of green manufacturing system, which increases the product life span and after PLC (product life cycle) the product can be recycled/re-manufactured easily. Xu et al. (2013) conducted multiple comparative studies on green supply chain practices, and the results show that green manufacturing increased environmental sustainability but also enhanced economic benefits, while green manufacturing practices cannot be implemented successfully or green manufacturing's vital objective is difficult to achieve without information sharing and collaboration through efficient and effective green information system. Xu et al. (2013) argue that electronic firms have lower pressures of GSCM practices as compared to the chemical and vehicle sector, while policymakers and senior management can get a good indicator of components that motivate a firm towards green manufacturing practices or environmental-friendly manufacturing. Therefore, we initiated the following hypothesis to measure the effect of GIS (green information systems) on GM (green manufacturing).

H-6: GIS (green information systems) have a significant and positive impact on GM (green manufacturing).

2.3.7 *Green Information System and Green Distribution*

During the implementation of green supply chain best practices, a green information system plays a vital role in sharing reliable information with suppliers and downstream levels. The green information system creates active collaboration and coordination among and within supply chain departments. The dream of green supply cannot become real without the proper adoption of green practices in the firm's distribution, while the distribution operation is more connected with the downstream level. The main objective of green distribution is to reduce the harmful effect on the environment and enhance customer service and product value while reducing inventory level, the enhanced firm image in the customers. The green supply chain activities not only affect the firm's operational and financial performance but also improved and build strong competitiveness in the long term (Lau, 2011).

Green information system increased information sharing in the end-to-end supply chain management, which has a significant and positive effect on the higher velocity and reduction in waste from the supply chain management. On the other hand, green distribution and warehousing also have an essential role in the product and customer service value addition in terms of postponement strategies, cross-docking, break-bulking, consolidation, and merging of products. Reduction of energy and polluted materials is the essential requirement for the green practices in the distribution and warehousing (Khan & Qianli, 2017). In the modern era, many firms used warehousing and distribution network for multiple purposes mainly including green packaging and labeling, reduction of inventory level, collection of defected and return products, receiving feedback from the customers, repairing of defective products, and reverse logistics (Rostamzadeh et al., 2015; Xu et al., 2013).

Many firms are implementing green practices in their distribution and warehousing network in order to increase environmental sustainability and enhance the firm's image and financial benefits and reduce footprints and pollution (Memari, Abdul Rahim, Absi, Ahmad, & Hassan, 2016). The following sub-criteria usually used for implementing green practices in the distribution network: Ecological packaging, decrease inventory levels, sale scrap, and excess inventories/materials, sale of excel capital equipment, used distribution centers for postponement strategy, cross-docking, break-bulking, consolidation, green packaging, and labeling, used distribution centers for reverse logistics (defective products and products return), collection of customer feedback, ensure rapid delivery of products to customers and reduced carbon emissions (Kim & Min, 2011). Accordingly, we build a hypothesis that GIS (green information system) has a significant impact on the improvement of GD (green distribution).

H-7: GIS (green information system) has a significant and positive effect on product GD (green distribution).

2.3.8 Green Information System and Green Transportation

Green information systems enhanced information sharing among supply chain stakeholders, which create a significant effect on the improvement of green transportation and the firm's performance. Carbon emission from transportation could seriously damage the fauna and flora and human lives. The transportation sector accounts for almost 15% of overall GHG (greenhouse gases) emissions and around 23% of carbon emissions. From 1990 to 2007, the global carbon emission from the transportation sector has grown by 45%, and green transportation has a significant role in reducing carbon emission and improving environmental sustainability around the globe. Even though the transportation role in serving green transportation agenda is comparatively ignored but its fact that the firm cannot ignore the part of green transportation in the implementation of green supply chain projects (Geng et al., 2013).

Generally, a green information system is developed if the senior leadership is well-committed towards the green supply chain project and green IS increased to the information, monitoring of the green project, and also focused on environmental performance evaluation. The primary objective of the green supply chain is to improve environmental sustainability and enhance economic benefits and create a competitive edge with positive image building in the customer mind. On the other hand, green transportation is mainly working on the reduction of carbon emissions, fossil fuel consumption, and increased renewable energy sources to promote environmental sustainability (Degirmenci & Breitner, 2017). In the last couple of years, researchers pay attention to the role of transportation play in achieving positive environmental performance. Green transportation can smash with the designated financial goals. Exploring the association between environmental effects and transportation by the planning of route will be able to offer valuable and practical suggestions regarding green supply chain management (Lin, Choy, Ho, Chung, & Lam, 2014). Ronald (2016) researched green supply chain management in Indian dairy firms. The results show that the adoption of sophisticated green practices in transportation not only reduces the harmful effect on the environment but also increases economic benefit and build a competitive edge for the firms. The vehicles with advanced green technology (Ronald, 2016), such as using more aerodynamic trucks and more fuel-efficient/fuel-powered trucks, are crucially recommended for the green supply chain projects (Lin et al., 2014).

Green transportation system enhanced and positive impact on environmental sustainability in terms of reduced carbon emissions, control climate change issues, and global warming/temperature. The green transportation mainly covers: energy-efficient vehicles, proper maintenance of the vehicles, provide green training to the transport-related employees, encouraging driving to reduce fossil fuel consumption, encourage renewable energy/green fuel sources such as liquid natural gas and biofuels (Rostamzadeh et al., 2015). Therefore, we initiated the following hypothesis to measure the effect of GIS (green information systems) on GT (green transportation).

H-8: GIS (green information systems) have a significant and positive impact on GT (green transportation).

2.3.9 Green Information System and Ecological Design

Ecological design plays a vital role in the adoption of green supply chain management projects, while the green information system is the motivating factor of ecological design. Product stewardship incorporates in the situation of external members into the ecological development stages through product life cycle analysis; green products can measure the product's effect on environmental sustainability Green et al., 2012). Firms also build a competitive edge through the adoption of ecological design and enhance environmental and economic benefits and also develop unique manufacturing capabilities. Gronhaug and Kaufmann

(1988) conducted a research and found that eco-friendly product has a significant and positive effect on the firm's survival and plays a significant role to create a sustainable competitive edge in the international market. Reinhardt (1998) green product design and competitive edge, the researcher found that eco-friendly products and eco-friendly practices into firms' product design process gain a competitive edge and also differentiate to the firm from their competitors.

Undeniably, green design and environmental-friendly practices are mainly dependent on (Khan & Qianli, 2017) customer requirements, supplier and customer management, and senior leadership commitment. Büyüközkan and çifçi, G. (2012) claimed that firms might reduce more than 80% of their negative and harmful effect from their green products and processes on environmental sustainability due to successful adoption of ecological practices in the product design phase. Gonzalez-Benuito and Gonzalez-Benito (2005) argue that companies emphasize ecological design which may reduce the usage of toxic and hazardous chemicals in their product and also minimize resource consumption in the production of goods. Zhu et al. (2007) conducted their research on Chinese firms and found that ecological design considerably reduced harmful effect on environmental sustainability and enhanced customer service with better firm's image and export opportunities.

Ecological design facilitates firms in the re-manufacturing, reusability, recyclability, and disassembly of products, while green products have a direct effect on firms' environmental and economic performance (Khan & Qianli, 2017). Customers prefer to buy green products to attain the benefits of environmental protection, and firms reduced their cost through the adoption of green design approaches. Researched eco-design products; the findings indicate that ecological design has a direct and positive association with company environmental performance and economic performance. Besides, firms also build a competitive edge, and the implementation of green design practices significantly enhanced company image and reputation in the international arena. Green product design reduces the cost of goods and mainly improved the value of the product (Porter and van der Linde 1995).

GIS offers the information needed for collaboration and coordination with downstream and upstream in terms of green design, labeling, and packaging. Information sharing, through the use of GIS, is the leading enabler for supply chain management/sustainable supply chain management in order integration, collaboration, and coordination (Chandra et al., 2007). GIS also provides the necessary information, which helps to make decisions about green design in terms of energy and material consumption, recycling, reuse, recovery of materials, and re-manufacturing of products. Besides, GIS offers the information required to recover the firm's investment in scrap, excess inventories of material, and finished goods. Accordingly, we build a hypothesis that GIS (green information system) has a positive and significant effect on ED (ecological design of products).

H-9: GIS (green information system) has a significant and positive effect on product ED (ecological design of products).

2.3.10 Green Manufacturing System and Economic Performance

Green supply chain management project's main idea is to implement or adopt green practices in the firm to reduce carbon emissions, climate change, and global warming while increasing environmental and economic performance with higher customer service level (Luthra, Govindan, Kannan, Mangla, & Garg, 2017). Several firms are adopting green practices to improve their financial statements and image the international market and also grab maximum export opportunities. In general speaking, green manufacturing increased the efficiency of using raw materials and components with the least consumption of hazardous materials and toxic chemicals. Besides, increased environmental sustainability reduces environmental pollution such as carbon emissions, GHG (greenhouse gases), and global warming issues (Scur & Barbosa, 2017).

Green manufacturing processes are very flexible and also supportive of re-manufacturing, recycling, refurbishment of products after PLC (product life cycle) or PSL (product shelve life). The firms have the plan to implement green projects in their end-to-end supply chain management and need to mainly focus on green manufacturing practices, which are called the heart of GSCM (green supply chain management) . These firms conduct preventive maintenance and predictive maintenance, emphasizing on their product performance and enhancing their PLC (Scur & Barbosa, 2017); they also conduct disassembly of the finished goods for subsequent reuse/re-manufacturing/recycling (Sinha & Anand, 2017; Xu et al., 2013).

Green manufacturing mainly covers material exploitation, production planning, control activities, waste reduction in terms of material and energy consumption, processing time, waiting time (Chuang, 2014), setup time, etc. and also helps firm to achieve their financial goals (Liu et al., 2012). Research on green supply chain impact on firm competitive edge and performance in the context of China, the results indicate that firms need to adopt green manufacturing practices to improve their financial performance and competitive edge with environmental performance (Liu et al., 2012). Chinese consumers are also willing to buy green products, and that is why Chinese manufacturers are now focusing on GSCM and especially emphasizing implementing green manufacturing as a pilot project. However, there are electronic firms, which are not willing to cuddle GSCM that efficiently and effectively, as they already have long-term relationship with the targeted customer base and required no green manufacturing over to win or retain their customers (Ronald, 2016).

Firms minimized their waste and cost from the system by adopting green processes and technologies (Khan, Zhang, Anees, et al., 2018; Golpîra & Khan, 2019). Lean manufacturing and green manufacturing both are working, in the same way, to control and reduce waste from the manufacturing system, eliminate unnecessary processing steps, and increase the efficiency of manufacturing along with an improved company image and reputation in the domestic and international arena (Zhang et al., 2018a, b; Khan, Zaman, & Zhang, 2016). Furthermore, firms in the

European region have incorporated and integrated total quality management concepts and lean practices with green and environmental-friendly to attain greater environmental sustainability with the financial objective (Golpîra et al., 2018; Khan, 2019; Khan et al., 2019). Many types of research conducted in the automobile industry and results indicate that green manufacturing has a positive impact on a firm's competitive edge and green manufacturing increased the firm's performance (Basu & Corporation, 2011). Baines et al. (2012) argue that manufacturing with green practices plays a crucial role in successful implementation of green project to reduce the harmful effects on environmental sustainability with little waste in overall manufacturing systems and also support firms to increased efficiency in the processes, increasing financial benefits, customer satisfaction, brand image, and environmental performance supported by extended literature (Chen et al., 2006; Deif 2011; Farish 2009; Khan & Qianli, 2017; Luthra et al., 2017; Zailani et al. 2015). Accordingly, we build a hypothesis that GM (green manufacturing) positively and significantly enhanced ECP (economic performance).

H-10: GM (green manufacturing) positively and significantly enhanced ECP (economic performance of the firms).

2.3.11 Green Manufacturing System and Competitive Edge

Green manufacturing provides firms an opportunity to differentiate their services from other manufacturing firms and build a competitive edge. Liu et al. (2012) researched to test the correlation between green manufacturing and competitive advantage in the context of Chinese manufacturing firms. The results show that green manufacturing has a significant and positive effect on the firms' competitive edge. Green manufacturing practices mainly reduced waste from the processes in terms of unnecessary processes, reduced processing time, and setup time. Companies are become more reliant on the vendors and key suppliers to achieve a competitive advantage. Senior leadership is increasingly emphasized to deal with environmental and social issues to develop a competitive edge, and positive brand image (Laari, Tyli, & Ojala, 2017) companies might be held accountable not only for their individual but also for the social performance and environmental performance of their suppliers. Porter (1980) argues that concurrent chase of differentiation and cost leadership is unlikely to manufacture a green competitive edge, and the company would risk being "stuck in the middle." The largest manufacturing companies could be characterized as marketing differentiators whose competitive edge is mainly based on the product image and firm reputation. As a response to pressures from many supply chain members/stakeholders, such as customers, regulatory authority/government, NGOs (nongovernment organizations), and competitors (Seuring and Müller, 2008), companies have introduced supplier evaluation schemes, which integrate social and environmental criteria.

On the other hand, there are many strategic drivers and motivational factors to adopt green supply chain management, such as improved corporate image, increased efficiency, increased innovation leadership, enhanced firm reputation in the international market, and also trust development Laari et al., (2017). Properly implementation of green manufacturing practices may create innovations that lower the products' total cost or enhance its value in terms of customer satisfaction (Brandenburg, 2015), which creates a competitive advantage in the market and also increased firm market share and size. Furthermore, green manufacturing is a suitable and appropriate approach to reduce environmental risk and enhance corporate social responsibility. On the other hand, environmental incidents may strengthen government and regulatory pressures, damage the company' reputation and image, and devaluate firm's share price value in the stock exchange, and customers may stop buying/boycott the company's products and cancel their orders (Hajmohammad et al., 2013). Therefore, we develop the following hypothesis to measure the effect of GM (green manufacturing) on firms' CA (competitive edge of the firms).

H-11: GM (green manufacturing) has a significant and positive impact on firms' CA (competitive edge of the firms).

2.3.12 Green Manufacturing System and Environmental Performance

Green manufacturing practices increase environmental sustainability in the region, which also plays a vital role in attracting foreign investors to invest in the country. Mainly the green manufacturing deals with the adoption of environmental-friendly means while implementing many actions in the progression of an SC (supply chain). The green supply chain management is a long term that covers within itself a complete concept including green manufacturing, material management (Ronald, 2016), green purchasing, green marketing, reverse logistics, green distribution, and waste management. In the traditional supply chain, manufacturing practices were using dangerous, hazardous, and toxic chemicals, which not only destroy the environmental sustainability but also create a significant harmful effect on the flora, fauna, and human lives (Dadhich, Genovese, Kumar, & Acquaye, 2015).

Green manufacturing minimizes harmful environmental effect by selecting products and goods which have been re-manufactured/recycled and refurbished, while green manufacturing practices create maximum flexibility in the manufacturing processes for recycling and re-manufacturing of products, which not only save money (Dadhich et al., 2015) but also enhanced environmental sustainability in the long run. Khan et al. (2017) explained that green supply chain management theory is mainly based on green manufacturing and supply chain technology. The aim is to reduce the negative environmental effect and enhance the primary objective of firm performance through the actions of the supply chain, management strategies, and the collaborative partnership (Chavez, Yu, Feng, & Wiengarten, 2016).

Geffen and Rothenberg (2000) conducted empirical research, and results show that in the manufacturing setting, secure and close relationships and collaboration with customers and suppliers lead to higher environmental performance. Green manufacturing practices mainly reduce the waste from the processes, reduce hazardous and toxic chemicals in the manufacturing of products, reduce fossil fuel and energy consumption, and promote the usage of renewable/green energy sources and biofuels to reduce carbon emission, climate change, and global warming in the region. According to the anecdotal evidence, Green et al. (1998) argue that supply policies and green manufacturing practices are likely to indicate in enhanced environmental performance. The bottom line of the green manufacturing practices is to promote green practices in the supply chain and reduce environmental degradation and also increased environmental performance. Accordingly, we build a hypothesis that GM (green manufacturing) positively and significantly enhanced ENP (environmental performance) of the firms.

H-12: GM (green manufacturing) positively and significantly enhanced ENP (environmental performance of the firms).

2.3.13 Green Distribution and Competitive Advantage

Green distribution practices create a positive image in the customers' market and fulfill the requirement of their customers to increase environmental sustainability, which creates a competitive edge for the firm to survival in the long run. Liu et al. (2012) researched the impact of green supply chain and distribution practices to increase competitive edge. The results show that competitive edge builds by the adoption of green distribution in the end-to-end supply chain. On the other hand, green supply chain management is a broader name of green concept to enhance environmental sustainability without creating a negative effect on the substantial financial benefits (Kushwaha & Sharma, 2016).

Many types of research are conducted to analyze the relationship between green distribution practices and competitive advantage, and most researchers found that green distribution is not only correlated with competitive advantage but also has a significant and positive effect on the financial and environmental performance of the firms (Khan & Qianli, 2017; Memari et al., 2016). Since the last couple of decades, the distribution center is not only used for storing goods and inventory but also used for product value addition such as postponement strategies, cross-docking, break-bulking, consolidation, reverse logistics (returned products, repairing, recycling), green packaging, and labeling (Accorsi, Cascini, Cholette, Manzini, & Mora, 2014).

In the last couple of years, many companies are adopting green practices in their supply chain to improve the firm's performance, and companies have understood the criticality of environmental-friendly distribution network importance. However, the starting cost and time investment to convert to this kind of distribution center terrify various companies. To enhance corporate social responsibility and reduce distribution costs, many distributions and fulfillment firms are looking to adopt green practices that reduce carbon dioxide emission, carbon footprint, and environmental pollution. The following criteria are mainly used to adopt green practices in the distribution network: green packaging and labeling, inventory level reductions, investment recovery, sale of unnecessary materials, excess inventory and scrap, etc. (Liu et al. 2012). The findings show that green distribution has no such direct effect on firms' performance but has significantly and positively affected the competitive building edge, which enhances overall company performance. For that reason, the best practice companies should emphasize on the downstream (customers and distribution) and upstream (suppliers) of green supply chain management. Therefore, we build the following hypothesis to measure the effect of GD (green distribution) on firms' CA (competitive edge of the firms).

H-13: GD (green distribution) has a significant and positive effect on firms' CA (competitive edge).

2.3.14 Green Distribution and Environmental Performance

Green distribution has a critical role in the adoption of green supply chain management, and green distribution has a direct effect on the enhancement of environmental performance. (Garofalo, Andrea, Tomaiuolo, Venezia, & Castrignan, 2017) the perspective of Italian food products, the results shows that green distribution is the essential factors to enhanced firm image in the customer mind, increased customer satisfaction, improved firm image and reputation, support of government and regulatory authorities, ethical support of NGOs (nongovernment organizations) and international community, attractive export opportunities, more significant financial and environmental advantage (Walmsley, Liu, Varbanov, & Klemeš, 2015).

Carbon emissions, climate change, global warming, and global temperature may be controlled by the adoption of green practices in the distribution networks (Liotta, Stecca, & Kaihara, 2015). Because of the substantial movement of transportation required in the distribution center and warehouse in terms of picking, storing, sorting packaging, assortment, break-bulking, cross-docking, and consolidation of products (Ryoo & Koo, 2013). Countries' image also increased by the adoption of environmental practices in the business and supply chain operations, while foreign direct investment inflows and per capita income are correlated with green practices (Bi, Wang, & Yang, 2013). On the other hand, the tourism industry is also booming due to the implementation of environmental-friendly practices and strict environmental policies implemented by the federal and provincial governments (Fiore, Silvestri, Contò, & Pellegrini, 2016). In the perspective of Pakistan, during the last 4 years, the provincial government has done plantation more than one billion trees in the province, which increased environmental sustainability but also attracted foreign direct investment inflows and foreign tourist. Climate change and global warming only can control by practical action taken by regulatory authorities and firms as corporate social responsibility (Villanueva-Ponce et al., 2015). Accordingly, we build a hypothesis that GM (green manufacturing) positively and significantly enhanced ECP (economic performance).

H-14: GD (green distribution) is positively and significantly improved ENP (environmental performance of the firms).

2.3.15 Green Transportation and Competitive Advantage

In the adoption of green supply chain management, green transportation received minimal importance, but in real case, green transportation has significant contribution in the reduction of carbon emission, greenhouse gases, and global warming and controlling climate change and temperature rise (Kumar, Luthra, Govindan, Kumar, & Haleem, 2016). Green transportation helps and provides an opportunity to create a competitive edge through adoption of the best green practices, while customers are more willing to buy firms' product if they know and physically assess firm transportation operations. Because transportation operation is straightforward to judge by customers and the firm's vehicle itself creates positive image building and reputation (Abbasi & Nilsson, 2016).

The fatality gas emissions resulted from vehicles have never been a mystery for companies. The transport sector emits 15% GHG (greenhouse gases) emission and 23% of carbon dioxide emissions from their activities and movement (Yang et al., 2013). Since 1990 to 2007, a growth of 45% in carbon dioxide emissions in the world is recorded, and it is continuously increasing and it is expected that from 2007 to 2030 carbon emission will rise to 45%, which will create environmental unsustainability, climate change, global warming, and temperature rise, causing humans, flora, and fauna to suffer in the long run (Rostamzadeh et al., 2015).

Green transportation is a significant practice in the project of green supply chain management. Green transportation's objective is to increase environmental sustainability and control carbon, GHG emissions, and global warming. There are many factors includes in green transportation but mainly are; reduction of fossil fuel consumption, shifting vehicles on renewable/green energy/electrical/solar energy, proper maintenance of vehicles on periodically basis, providing suitable green training to the transport-related employees, use suitable and short routes to arrive destination/customers, consolidated delivery to customers, and buy fuel-efficient and advanced technology vehicles, which can not only reduce cost from the transportation system but also have significant improvement in environmental performance and grab competitive edge in terms of image and reputation building. Therefore, we build the following hypothesis to measure the correlation between GT (green transportation) and firms' CA (competitive edge).

H-15: GT (green transportation) has a significant and positive effect on firms' CA (competitive edge).

2.3.16 Green Transportation and Environmental Performance

The fundamental objective of environmental-friendly transportation is to reduce the harmful effect of the supply chain and business operations on environmental sustainability while enhancing customer satisfaction and environmental performance with building a competitive advantage. Ronald (2016) conducted a research on transportation which involved green practices and found that green transportation has a significant and positive effect on the environmental performance of the firms. The adoption of advanced green transportation dealing, which has less relation to cost savings and efficiency. Advanced green transportation has a vital role in the reduction of carbon emissions, climate change, and global warming (Yang et al., 2013).

Firms can control their business operations' harmful effect on environmental sustainability through the adoption of green transportation practices, which are based on aerodynamic vehicles, and more fuel (Synergies, Sun, Sabbaghi, & Ashton, 2017) powered vehicles are all adopted in terms to control their emissions and create positive image, reputation building and also grab maximum opportunities in the international arena. The environmental performance mainly relates to the ability of production firms to minimize carbon emissions, and solid wastes, and the ability to reduce consumption of toxic chemicals and hazardous materials (Green et al., 2012). On the other hand, green transportation cover and based on the reduction of fossil fuel consumption, reduction of energy and waste from the systems, usage of modern and environmental-efficient vehicles, using green and renewable fuels such as low sulfur content and biofuels, providing proper green environmental training to the relevant employees, etc. Accordingly, we build a hypothesis that GT (green transportation) positively and significantly improved ENP (environmental performance).

H-16: GT (green transportation) positively and significantly improved ENP (environmental performance of the firms).

2.3.17 Ecological Design and Economic Performance

Ecological design or green design of products directly reduces the cost of the products in terms of recycling, reuse, re-manufacturing, refurbishment, and recovery of materials and components from the product after product life cycle (PLC) (Alaharja & Helo, 2014). The first-mover benefits are where the company introduces green product design and enhances advantages covering getting royalties for licensing of green technology or products, unique production capabilities, and also create proprietary information which will lead to a sustainable competitive edge (Yang et al., 2013).

Gronhaug and Kaufmann (1988) describe that the green product design has a significant impact on a firm's survival and plays a vital and essential role of weapon to enhance environmental sustainability and create a sustainable competitive edge. Many firms are using green practices into their (Reinhardt 1998) product innovation in order to create competitive advantage and also differentiate themselves from other companies/competitors. Undeniably, ecological design practices are mainly dependent on supplier and customer management and senior leadership commitment towards green supply chain management practices. Büyüközkan and çifçi, G. (2012) confirmed that firms might reduce more than 80% harmful effect from their products and processes on the environment sustainability due to the successful implementation of ecological design practices. An empirical study on vehicle firms in China. Zhu et al. (2007) found that ecological design positively and significantly minimized the negative effect on environmental sustainability in terms of recycling, reuse, re-manufacturing, recovery materials, and components, reduced hazardous chemicals, and also increased company's contribution towards corporate social responsibility.

Gonzalez-Benuito and Gonzalez-Benito (2005) argue that the company emphasizes that ecological design may reduce the usage of toxic and hazardous materials in products and also may minimize resource consumption in the production of products. Besides, environmental-friendly product design also facilitates recycling, re-manufacturing, reusable, and recovery of components and materials from products after product life cycle. The green design of products has a significant effect on firm performance, and similarly, customers are more willing to purchase green products to attain the benefits of environmental protection and cost. The findings indicate that ecological design/green product design has a positive relationship with firm performance and creates a positive image and competitive edge and implementation of green design/environmental-friendly practices increased company share market. Further, eco-friendly and green products (Porter and van der Linde 1995) reduce the cost of products and meaningfully improve the value of the product. Accordingly, we build the following hypothesis to measure product ED (ecological design) effect on firm ECP (economic performance).

H-17: ED (ecological design) has a significant and positive effect on firms' ECP (economic performance).

2.3.18 Ecological Design and Competitive Advantage

Ecological product design creates a coma competitive edge for the firms in terms of positive brand image and improved firms' reputation. Ecological product design needs that producers design to the product which may reduce consumption of energy, materials, and components that facilitate to the firms in terms of recovery of materials and parts, reuse, recycle and re-manufacturing, or refurbishment (Zhu et al., 2008a).

Diabat and Govindan (2011), the eco-friendly product designs and their impacts on the performance, found that many motivating factors create a direct and significant effect on the adoption of green supply chain practiced and found that ecological design has a positive influence on the firms' performance. The principal objective of ecological design is to minimize products' harmful effects on the environment without reduction of firms' internal profitability. Khan and Dong (2017b) the green supply chain practices impact on the manufacturing firms, they found that ecological design not only reduces harmful effect on environmental sustainability but also enhanced firms' performance in terms of recycling, re-manufacturing, and reuse approaches, the further the green design has a positive and significant impact on the organizational performance. Gabriel and Benito, (2005) argues that organizations focus on environmental-friendly design may reduce the usage of hazardous materials and chemicals from the products and also minimize resource consumption in production goods. Furthermore, ecological design products enhance enterprise performance, and customers are also satisfied with firms, which create a coma competitive edge for the firms in terms of positive brand image and reputation in the domestic market and international markets. Besides, eco-friendly design products create uncountable export opportunities for firms. Accordingly, we build a hypothesis that ED (ecological design of products) positively and significantly creates CA (competitive edge for the manufacturing firms).

H-18: ED (ecological product design) positively and significantly creates CA (competitive edge for the companies).

2.3.19 *Ecological Design and Environmental Performance*

Green product design is a vital part of the concept of green supply chain management; ecological design reduces environmental pollution in terms of carbon dioxide, greenhouse gas emission, global warming, climate (Suburban, Taşan, & Baykasoğlu, 2015) change, and world temperature. The manufacturing firms create several problems for the environmental sustainability and increased air pollution and water pollution, while ecological design talks about environmental sustainability and the green environment with improved firms' performance and creates a competitive edge in terms of positive image building and reputation (Khan & Qianli, 2017).

We claimed that green design and eco-friendly product design increased customer satisfaction and environmental sustainability, and that have been manufactured by processes that operated and designed for the improved environment (Nie, 2016). The green product design provides an opportunity to save their cost in terms of recycling, re-manufacturing, reuse, and refurbishment, while green product design enhanced the environmental performance of the (Green et al., 2012) firms with financial benefits. Accordingly, we build a hypothesis that ED (ecological design of products) positively and significantly enhanced ENP (environmental performance of the manufacturing firms).

H-19: ED (ecological product design) positively and significantly enhanced the ENP (environmental performance of the companies).

2.3.20 *Environmental Performance and Economic Performance*

The adoption of green supply chain management practices directly increased environmental performance and sustainability but also enhance financial performance indirectly (Khan & Qianli, 2017) green supply chain focus on waste elimination related to environmental sustainability. Conducted a research on the environmental-friendly supply chain and its effect on economic performance. The results show that the association between GSCM and economic performance is influential, and this relationship also creates a significant effect on the competitive edge (Hayami, Nakamura, & Nakamura, 2015).

Generally, economic performance refers to the production plant's ability to minimize costs related to consumption of energy, purchased materials and components, waste discharge, waste treatment, and heavy penalties and fine for environmental accidents (Zhu et al., 2008a).

The cost-saving nature of environmental performance needs to achieve and enhance financial performance and build competitive edge with positive image building and firms' reputation in the international market (Fallahpour, Udoncy Olugu, Nurmaya Musa, Yew Wong, & Noori, 2017), while financial performance also needs to enhance its operational performance and reflect firms' ability to satisfy customers' demand for green products and services. The marketing implications of environmental performance lead to a significant enhancement in economic performance (Kara, Ibbotson, & Kayis, 2014). Accordingly, we build a hypothesis that ENP (environmental performance of the firms) positively and significantly enhanced ECP (economic performance of the manufacturing firms).

H-20: ENP (environmental performance) positively and significantly enhanced ECP (economic performance).

2.3.21 Environmental Performance and Organizational Performance

Undeniably, environmentally friendly practices increased the positive effect on the environmental performance of the firms, and it also enhanced organizational performance. The saving of cost nature of environmental performance should attain to increase both economic and operational performance with overall organizational performance (Green et al., 2012). Green practices implementation in the supply chain generate cost-saving and echo a firm's ability to fulfill the demand of customers for green products and services. The marketing and cost implication of operational and environmental performance may lead to a significant contribution to firms' overall performance (Van Hoof, 2014).

Green supply chain management practices and manufacturing firms in the context of the US Environment do not directly impact firms' performance; slightly, they indirectly affect on company (Wei, Wong, & Lai, 2012) performance through better green practices (Green et al. 2012). Finally, the association among the performance constructs seems to link logical sense. Green environmental performance leverage improved operational performance, which leads to increased overall firms' performance (Vranakis & Chatzoglou, 2014). Production managers have had to build robust knowledge and skills of green supply chain practices to manage at the firm level. Top management must emphasize enhancing SC in terms of enhancing overall firm performance. The primary source of income for firms is customers; if customers create pressure on firms to produce green products, ultimately manufacturing firms will produce those products to sustain their final customers, and it will also create positive effect on the environmental and organizational performance of the companies (Weerawardena, Mort, Salunke, Knight, & Liesch, 2014).

Once green practices have become strategic attention and information systems have been modified to monitor and measure the environmental efforts of firms, manufacturers can begin to adopt green practices with some extent of confidence that the practices will yield performance but also enhanced overall company performance (Luna-Reyes et al., 2014). The execution of green supply chain management practices enhances the overall companies' capabilities to sustain the environment and to support the overall firms' performance. Accordingly, we build a hypothesis that ENP (environmental performance) positively and significantly enhanced ORP (organizational performance of the manufacturing firms).

H-21: ENP (environmental performance) positively and significantly enhanced ORP (organizational performance).

2.3.22 *Competitive Advantage and Organizational Performance*

Green supply chain management practices create a competitive advantage in the domestic and international markets, while competitive edge creates a significant and positive effect on the overall organizational performance. Liu et al. (2012) conducted the research on green supply chain practices and competitive edge, and the results show that competitive advantage build by green practices including green manufacturing, green design/ecological product design, green transportation, green warehousing, and green material management (Kara et al., 2014) which create significant and direct impact on the building competitive advantage for the firms and ultimately competitive advantage enhanced firms' overall performance (Scur & Barbosa, 2017).

Organizational performance refers to the marketing performance and financial performance of the firms as compared to the industry average firms' performance. Green supply chain practices in the of US manufacturing firms, the findings confirmed that green practices enhance environmental performance and build competitive advantage in the market as compared to their competitors, which competitive advantage finally, enhanced to the overall organizational performance/firm performance (Afif et al., 2013; Lau, 2011). Therefore, we build a hypothesis that CA (competitive advantage) positively and significantly enhanced ORP (organizational performance of the manufacturing firms).

H-22: CA (competitive advantage) positively and significantly enhanced ORP (organizational performance of the manufacturing firms).

2.3.23 Economic Performance and Organizational Performance

The adoption of green supply chain management practices mainly including green purchasing, ecological design, cooperation with customers and suppliers, green manufacturing, green distribution, and transportation leads to higher economic performance and competitive edge, while the economic performance of the firms enhanced overall organizational performance. Green et al., (2012) the results indicate that economic performance increased by the adoption of green practices in the supply chain management, while organizational performance affected/influenced by higher economic performance (Cohen & Roucsel, 2005).

The green technology investment enhances environmental performance but also increases financial performance. Besides, the economic performance of the firms leads to enhanced organizational performance (Rao and Holt 2005). Many companies such as Engro Corporation, General Motors, and Protector & Gamble invest $3 billion to implement advanced technology and green practices to improve their environmental and economic performance (Khan & Qianli, 2017), because they know its economic benefits will directly and indirectly improve the company's overall performance in the long run. Therefore, we build a hypothesis that ECP (economic performance) positively and significantly enhanced ORP (organizational performance of the manufacturing firms).

H-23: ECP (economic performance) positively and significantly enhanced ORP (organizational performance of the manufacturing firms).

References

Abbasi & Nilsson (2016). Developing environmentally sustainable logistics: exploring themes and challenges from a logistics service providers' perspective, Transportation Research Part D: *Transport and Environment, 46*, 273–283.

Accorsi, R., Cascini, A., Cholette, S., Manzini, R., & Mora, C. (2014). Economic and environmental assessment of reusable plastic containers: A food catering supply chain case study. *International Journal of Production Economics, 152*, 88–101. https://doi.org/10.1016/j.ijpe.2013.12.014

Afif, A., Norlida, M., Hardi, J., Rosly, E., Abdul, H., Afif, A., Delopo, I. (2013). Article information : https://doi.org/10.1108/S1479-3563(2012)000012B005

Ala-harja, H., & Helo, P. (2014). Green supply chain decisions – Case-based performance analysis from the food industry. *Transportation Research Part E, 69*, 97–107. https://doi.org/10.1016/j.tre.2014.05.015

Baines, T., Brown, S, Benedettini, O., & Ball, P. (2012). Examining green production and its role within the competitive strategy of manufacturers. *Journal of Industrial Engineering and Management, 5*(1), 53–87.

Bakhodirov, B., & Peng, Q. (2009). *E c o n o m y diffusion mechanism and models of green supply chain management* (pp. 131–145).

Basu, S., & Corporation, I. (2011). Looking towards a sustainable and green future.

Bi, K. X., Wang, Y. H., & Yang, C. J. (2013). Study on the influence of FDI on China's green innovation in manufacturing system. *Applied Mechanics and Materials, 448–453*, 4571–4576. https://doi.org/10.4028/www.scientific.net/AMM.448-453.4571

Bowen, T. J. (2001). The relationship between customer loyalty and customer satisfaction. *International Journal of Contemporary Hospitality Management, 13*(5), 213–217.

Brandenburg, M. (2015). Low carbon supply chain configuration for a new product – A goal programming approach. *International Journal of Production Research, 7543*(April), 1–23. https://doi.org/10.1080/00207543.2015.1005761

Chandra, R. P., Bura, R., Mabee, W. E., Berlin, A., Pan, X., & Saddler, J. N. (2007). Substrate pretreatment: the key to effective enzymatic hydrolysis of lignocellulosics. *Adv Biochem Eng Biotechnol, 108*, 67–93.

Chavez, R., Yu, W., Feng, M., & Wiengarten, F. (2016). The effect of customer-centric green supply chain management on operational performance and customer satisfaction. *Business Strategy and the Environment, 25*(3), 205–220. https://doi.org/10.1002/bse.1868

Chen, Y.-S., Lai, S.-B., & Wen, C.-T. (2006). The Influence of Green Innovation Performance on Corporate Advantage in Taiwan. *Journal of Business Ethics, 67*(4), 331–339. https://doi.org/10.1007/s10551-006-9025-5

Chuang, S. P. (2014). Assessing and improving green performance using a compound approach. *Flexible Services and Manufacturing Journal, 26*(1–2), 69–91. https://doi.org/10.1007/s10696-012-9156-1

Cohen & Roucsel, (2005). Strategic supply chain, McGraw Hill Professional, ISBN 0071454497, 9780071454490, New York.

Dadhich, P., Genovese, A., Kumar, N., & Acquaye, A. (2015). Developing sustainable supply chains in the UK construction industry: A case study. *International Journal of Production Economics, 164*, 271–284. https://doi.org/10.1016/j.ijpe.2014.12.012

Degirmenci, K., & Breitner, M. H. (2017). Consumer purchase intentions for electric vehicles: Is green more important than price and range? *Transportation Research Part D: Transport and Environment, 51*, 250–260. https://doi.org/10.1016/j.trd.2017.01.001

Deif, A. M. (2011). A system model for green manufacturing. *Journal of Cleaner Production, 19*, 1553–1559.

Demir, E., Bekta, T., & Laporte, G. (2014). A review of recent research on green road freight transportation. *European Journal of Operational Research, 237*(3), 775–793. https://doi.org/10.1016/j.ejor.2013.12.033

Diabat & Govindan. (2011). An analysis of the drivers affecting the implementation of green supply chain management. *Resources, Conservation and Recycling, 55*(6), 659–667.

Dongmin, X., Dingyi, Youfang, H., & Bin, Y. (2016). A study of cost and emissions considering speed optimization in trans Eurasia transportation based on intelligent transportation. *International Journal on Smart Sensing and Intelligent Systems, 9*(1), 191–214.

El-Gayar, O., & Fritz, B. (2006). Environmental Management Information Systems (EMIS) for Sustainable Development: A Conceptual Overview. Communications of the Association for Information Systems, 17, 11–34. https://doi.org/10.17705/1CAIS.01734.

Fallahpour, A., Udoncy Olugu, E., Nurmaya Musa, S., Yew Wong, K., & Noori, S. (2017). A decision support model for sustainable supplier selection in sustainable supply chain management. *Computers and Industrial Engineering, 105*, 391–410. https://doi.org/10.1016/j.cie.2017.01.005

Farish, M. (2009). Plants that are green: Toyota's lean manufacturing. *Engineering and Technology, 4*(3), 68–69.

Fiore, M., Silvestri, R., Contò, F., & Pellegrini, G. (2016). Understanding the relationship between green approach and marketing innovations tools in the wine sector. *Journal of Cleaner Production, 142*, 4085–4091. https://doi.org/10.1016/j.jclepro.2016.10.026

Gabriel, R., & Benito, G., (2005). Divestment and international business strategy. *Journal of Economic Geography, 5*(2), 235–251. https://doi.org/10.1093/jnlecg/lbh041

Garofalo, P., Andrea, L. D., Tomaiuolo, M., Venezia, A., & Castrignan, A. (2017). Environmental sustainability of agri-food supply chains in Italy: The case of the whole-peeled tomato production under life cycle assessment methodology., *200*, 1–12. https://doi.org/10.1016/j.jfoodeng.2016.12.007

Geffen, C. A., & Rothenberg, S. (2000). Suppliers and environmental innovation: the automotive paint process. *International Journal of Operations and Production Management, 20*(2), 166–186.

Geng, Y., Ma, Z., Xue, B., Ren, W., Liu, Z., & Fujita, T. (2013). Co-benefit evaluation for urban public transportation sector – a case of Shenyang, China. *Journal of Cleaner Production, 58*, 82–91. https://doi.org/10.1016/j.jclepro.2013.06.034

Golpîra, H., & Khan, S. A. R. (2019). A multi-objective risk-based robust optimization approach to energy management in smart residential buildings under combined demand and supply uncertainty. *Energy, 170*(3), 1113–1129.

Golpîra, H., Khan, S. A. R., & Zhang, Y. (2018). Robust smart energy efficient production planning for a general job-shop manufacturing system under combined demand and supply uncertainty in the presence of grid-connected microgrid. *Journal of Cleaner Production, 202*(11), 649–665.

Gonzalez-Benuito & Gonzalez-Benito. (2005). An analysis of the relationship between environmental motivations and ISO14001 certification. *British Academy of Management, 16*(2), 133–148.

Green, M. A., Aller, R. C., & Aller, J. Y. (1998). Influence of carbonate dissolution on survival of shell-bearing meiobenthos in nearshore sediments. *Limnology and Oceanography, 43*(1), 18–28.

Green, K. W., Zelbst, P. J., Meacham, J., Bhadauria, V. S., Kenneth, W., Jr, G., Bhadauria, V. S. (2012). Green supply chain management practices: Impact on performance. Supply Chain Management: An International Journal, 17(3), 290–305. https://doi.org/10.1108/13598541211227126

Gronhaug, K., & Kaufmann, G. (1988). Innovation: A Cross-Disciplinary Perspective, Norwegian University Press: Oslo.

Gualandris, J., & Kalchschmidt, M. (2015). Supply risk management and competitive advantage: a misfit model. *International Journal of Logistics Management, 26*(3), 459–478. https://doi.org/10.1108/IJLM-05-2013-0062.

Hajmohammad, S., Vachon, S., Klassen, R. D., & Gavronski, I. (2013). Reprint of lean management and supply management : Their role in green practices and performance q. *Journal of Cleaner Production, 56*, 86–93. https://doi.org/10.1016/j.jclepro.2013.06.038

Halldórsson, Á., & Kovács, G. (2010). The sustainable agenda and energy efficiency: Logistics solutions and supply chains in times of climate change. *International Journal of Physical Distribution & Logistics Management, 40*(1/2), 5–13. https://doi.org/10.1108/09600031011018019

Hayami, H., Nakamura, M., & Nakamura, A. O. (2015). Economic performance and supply chains: The impact of upstream firms' waste output on downstream firms' performance in Japan. *International Journal of Production Economics, 160*, 47–65. https://doi.org/10.1016/j.ijpe.2014.09.012

Hsu, C. W., & Hu, A. H. (2008). Green supply chain Management in the Electronic Industry. *International Journal of Science and Technology, 52*, 205–216.

Jabbour, A. B., Jabbour, C., Govindan, K., Kannan, D., & Arantes, A. F. (2014). Mixed methodology to analyze the relationship between the maturity of environmental management and the adoption of green supply chain management in Brazil. *Resources, Conservation and Recycling, 92*, 255–267. https://doi.org/10.1016/j.resconrec.2014.02.004

Jabbour, A. B. L. D. S., Frascareli, F. C. D. O., & Jabbour, C. J. C. (2015). Green supply chain management and firms' performance: Understanding potential relationships and the role of green sourcing and some other green practices. *Resources, Conservation and Recycling, 104*, 366–374. https://doi.org/10.1016/j.resconrec.2015.07.017

Kara, S., Ibbotson, S., & Kayis, B. (2014). Sustainable product development in practice: An international survey. *Journal of Manufacturing Technology Management, 25*(6), 848–872. https://doi.org/10.1108/JMTM-09-2012-0082

Khan, S. A. R. (2019). The nexus between carbon emissions, poverty, economic growth, and logistics operations-empirical evidence from southeast Asian countries. *Environmental Science and Pollution Research*. https://doi.org/10.1007/s11356-019-04829-4

Khan, S. A. R., Chen, J., Zhang, Y., Golpîra, H., Kumar, A., & Arshian, S. (2019). Environmental, social, and economic growth indicators spur logistics performance: From the perspective of the South Asian Association for Regional Cooperation countries. *Journal of Cleaner Production, 214*(3), 1011–1023.

Khan, S. A. R., & Dong, Q. (2017a). Impact of green supply chain management practices on firms' performance: An empirical study from the perspective of Pakistan. *Environmental Science and Pollution Research, 24*(20), 16829–16844.

Khan, S. A. R., & Dong, Q. (2017b). Does national scale economic and environmental indicators spur logistics performance? Evidence from UK. *Environmental Science and Pollution Research, 24*(34), 26692–26705.

Khan, S. A. R., Dong, Q., SongBo, W., Zaman, K., & Zhang, Y. (2017a). Environmental logistics performance indicators affecting per capita income and sectoral growth: Evidence from a panel of selected global ranked logistics countries. *Environmental Science and Pollution Research, 24*(2), 1518–1531.

Khan, S. A. R., Dong, Q., SongBo, W., Zaman, K., & Zhang, Y. (2017b). Travel and tourism competitiveness index: The impact of air transportation, railways transportation, travel, and transport services on international inbound and outbound tourism. *Journal of Air Transport Management, 58*(1), 125–134.

Khan, S. A. R., Dong, Q., & Zhang, Y. (2016). Research on the measuring performance of green supply chain management: In the perspective of China. *International Journal of Engineering Research in Africa, 27*(12), 167–178.

Khan, S. A. R., Dong, Q., & Zhang, Y. (2017). Role of ABC analysis in the process of efficient order fulfillment: A case study. *Advanced Engineering Forum, 23*(7), 114–121.

Khan, S. A. R., Dong, Q., & Zhang, Y. (2018). The impact of green supply chain practices in business performance: Evidence from Pakistani FMCG firms. *Journal of Advanced Manufacturing Systems, 17*(2), 267–275.

Khan, S. A. R., Dong, Q., Zhang, Y., & Khan, S. S. (2017). The impact of green supply chain on enterprise performance: In the perspective of China. *Journal of Advanced Manufacturing Systems, 16*(3), 263–273.

Khan, S. A. R., & Qianli, D. (2017). Impact of green supply chain management practices on firms' performance: An empirical study from the perspective of Pakistan. *Environmental Science and Pollution Research.* https://doi.org/10.1007/s11356-017-9172-5

Khan, S. A. R., Zaman, K., & Zhang. (2016). The relationship between energy-resource depletion, climate change, health resources, and the environmental Kuznets curve: Evidence from the panel of selected developed countries. *Renewable and Sustainable Energy Reviews, 62*(9), 468–477.

Khan, S. A. R., & Zhang, Y. (2019). *Strategic supply chain management* (EAI/Springer Innovations in Communication and Computing Series). Springer Publisher, ISBN 978-3-030-15058-7, https://www.springer.com/gp/book/9783030150570

Khan, S. A. R., Zhang, Y., Anees, M., Golpîra, H., Lahmar, A., & Dong, Q. (2018). Green supply chain management, economic growth, and environment: A GMM based evidence. *Journal of Cleaner Production, 185*(6), 588–599.

Khan, S. A. R., Zhang, Y., Kumar, A., Zavadskas, E., & Streimikiene, D. (2020) Measuring the impact of renewable energy, public health expenditure, logistics and environmental performance on sustainable economic growth, Sustainable Development, Online Version, https://doi.org/10.1002/sd.2034.

Kim, I., & Min, H. (2011). Measuring supply chain efficiency from a green perspective., *34*(11), 1169–1189. https://doi.org/10.1108/01409171111178738

Kumar, S., Luthra, S., Govindan, K., Kumar, N., & Haleem, A. (2016). Barriers in green lean six sigma product development process: An ISM approach. *Production Planning and Control, 27*(7–8), 604–620. https://doi.org/10.1080/09537287.2016.1165307

Kushwaha, G. S., & Sharma, N. K. (2016). Green initiatives: A step towards sustainable development and a firm's performance in the automobile industry. *Journal of Cleaner Production, 121*). Elsevier Ltd. https://doi.org/10.1016/j.jclepro.2015.07.072

Laari, S., Tyli, J., & Ojala, L. (2017). Supply chain perspective on competitive strategies and green supply chain management strategies. *Journal of Cleaner Production, 141*, 1303–1315. https://doi.org/10.1016/j.jclepro.2016.09.114

Lai, M. (2008). Technology readiness, internet self-efficacy and computing experience of professional accounting students. *Campus-Wide Information Systems, 25*(1): 18–29.

Lau, K. H. (2011). Benchmarking green logistics performance with a composite index., *18*(6), 873–896. https://doi.org/10.1108/14635771111180743

Lee, J., Guan, Z., Akbergenova, Y., & Littleton, J. T. (2013). Genetic analysis of synaptotagmin c2 domain specificity in regulating spontaneous and evoked neurotransmitter release. *J. Neurosci. 33*(1): 187–200.

Lin, C., Choy, K. L., Ho, G. T. S., Chung, S. H., & Lam, H. Y. (2014). Survey of green vehicle routing problem: Past and future trends. *Exp Syst Appl, 41*(4 PART 1), 1118–1138. https://doi.org/10.1016/j.eswa.2013.07.107

Liotta, G., Stecca, G., & Kaihara, T. (2015). Optimisation of freight flows and sourcing in sustainable production and transportation networks. *International Journal of Production Economics, 164*, 351–365. https://doi.org/10.1016/j.ijpe.2014.12.016

Liu, J. P., Curry, J. A., Wang, H., Song, M., & Horton, R. M. (2012). Impact of declining Arctic sea ice on winter snowfall. *Proc. Natl. Acad. Sci., 109*, 4074–4079. https://doi.org/10.1073/pnas.1114910109.

Luna-Reyes, L. F., Zhang, J., Whitmore, A., Jarman, H., Picazo-Vela, S., Andersen, D. L., et al. (2014). Full information product pricing: An information strategy for harnessing consumer choice to create a more sustainable world. *Communications of the Association for Information Systems, 34*(1), 637–654.

Luthra, S., Govindan, K., Kannan, D., Mangla, S. K., & Garg, C. P. (2017). An integrated framework for sustainable supplier selection and evaluation in supply chains. *Journal of Cleaner Production, 140*, 1686–1698. https://doi.org/10.1016/j.jclepro.2016.09.078

Martí, J. M. C., Tancrez, J.-S., & Seifert, R. W. (2015). Carbon footprint and responsiveness tradeoffs in a supply chain network design. *International Journal of Production Economics, 166*, 129–142. https://doi.org/10.1016/j.ijpe.2015.04.016

Memari, A., Abdul Rahim, A. R., Absi, N., Ahmad, R., & Hassan, A. (2016). Carbon-capped distribution planning: A JIT perspective. *Computers and Industrial Engineering, 97*, 111–127. https://doi.org/10.1016/j.cie.2016.04.015

Mohammed, F., Selim, S. Z., Hassan, A., & Syed, M. N. (2017). Multi-period planning of closed-loop supply chain with carbon policies under uncertainty. *Transportation Research Part D: Transport and Environment, 51*, 146–172. https://doi.org/10.1016/j.trd.2016.10.033

Nie, X. (2016). Research on the management innovation in the green supply chain: An empirical analysis based on probit model. *International Journal of Smart Home, 10*(2), 153–164. https://doi.org/10.14257/ijsh.2016.10.2.15

Orji, I. J., & Wei, S. (2015). Dynamic modeling of sustainable operation in a green manufacturing environment. *Journal of Manufacturing Technology Management, 26*(8), 1201–1217. https://doi.org/10.1108/JMTM-11-2014-0120

Porter, M. E. (1980). Competitive Strategy: Techniques for Analyzing Industries and Competitors. New York: Free Press.

Polonsky & Ottman (1998). Reevaluating green marketing: a strategic approach. *Business Horizons, 44*(5), 21–30.

Rao, P., & Holt, D. (2005). Do green supply chains lead to competitiveness and economic performance? *International Journal of Operations & Production Management, 25*(9), 898–916. https://doi.org/10.1108/01443570510613956.

Reinhardt, (1998). Environmental product differentiation: implications for corporate strategy. *California Management Review, 40*(4), 41–61. https://doi.org/10.2307/41165964

Ronald, O. (2016). Journal of manufacturing technology management. *Journal of Manufacturing Technology Management., 27*(7). https://doi.org/10.1108/JMTM-10-2016-0134

Rostamzadeh, R., Govindan, K., Esmaeili, A., & Sabaghi, M. (2015). Application of fuzzy VIKOR for evaluation of green supply chain management practices. *Ecological Indicators, 49*, 188–203. https://doi.org/10.1016/j.ecolind.2014.09.045

Ryoo, S. Y., & Koo, C. (2013). *Green practices-IS alignment and environmental performance : The mediating effects of coordination, 799–814*. https://doi.org/10.1007/s10796-013-9422-0

Scur, G., & Barbosa, M. E. (2017). Green supply chain management practices: Multiple case studies in the Brazilian home appliance industry. *Journal of Cleaner Production, 141*, 1293–1302. https://doi.org/10.1016/j.jclepro.2016.09.158

Seuring & Müller (2008). From a literature review to a conceptual framework for sustainable supply chain management. *Journal of Cleaner Production, 16*, 1699–1710.

Sharma, V. K., Chandna, P., & Bhardwaj, A. (2017). Green supply chain management related performance indicators in agro-industry: A review. *Journal of Cleaner Production, 141*, 1194–1208. https://doi.org/10.1016/j.jclepro.2016.09.103

Sinha, A. K., & Anand, A. (2017). Towards fuzzy preference relations based on a decision-making approach to access the performance of suppliers in the environmentally conscious manufacturing domain. *Computers & Industrial Engineering, 105*, 39–54. https://doi.org/10.1016/j.cie.2016.12.033

Suburban, K., Taşan, A. S., & Baykasoğlu, A. (2015). Designing an environmentally conscious tire closed-loop supply chain network with multiple recovery options using interactive fuzzy goal programming. *Applied Mathematical Modelling, 39*. https://doi.org/10.1016/j.apm.2014.11.004

Synergies, B., Sun, J., Sabbaghi, N., & Ashton, W. (2017). *Green Supply Chain Formation Through, 64*(1), 70–82.

Van Hoof, B. (2014). Organizational learning in cleaner production among Mexican supply networks. *Journal of Cleaner Production, 64*, 115–124. https://doi.org/10.1016/j.jclepro.2013.07.041

Villanueva-Ponce, R., Garcia-Alcaraz, J. L., Cortes-Robles, G., Romero-Gonzalez, J., Jimnez-Maas, E., & Blanco-Fernandez, J. (2015). Impact of suppliers? Green attributes in corporate image and financial profit: Case maquiladora industry. *International Journal of Advanced Manufacturing Technology, 80*(5–8), 1277–1296. https://doi.org/10.1007/s00170-015-7082-6

Vranakis, S., & Chatzoglou, P. (2014). Business factors related to manufacturing firms' performance. *Journal of Industrial Engineering and Management, 7*(1), 336–359. https://doi.org/10.3926/jiem.896

Walmsley, M. R. W., Liu, X., Varbanov, P. S., & Klemeš, J. J. (2015). Environmental footprint comparison amongst dairy, grain, and meat products in California. *Chemical Engineering Transactions, 43*, 109–114. https://doi.org/10.3303/CET1543019

Weerawardena, J., Mort, G. S., Salunke, S., Knight, G., & Liesch, P. W. (2014). The role of the market sub-system and the socio-technical sub-system in innovation and firm performance: A dynamic capabilities approach. *Journal of the Academy of Marketing Science*. https://doi.org/10.1007/s11747-014-0382-9

Wei, H. L., Ju, P. H., & Angkasa, Y. A. (2016). Implementing green supply chain management to achieve competitive advantage. *2016 5th IIAI International Congress on Advanced Applied Informatics (IIAI-AAI)* (pp 755–758). https://doi.org/10.1109/IIAI-AAI.2016.242

Welty & Becerra-Fernandez, (2001). Managing trust and commitment in collaborative supply chain relationships. *Communications of the ACM, 44*(6) 49–58, https://doi.org/10.1145/376134.376170.

Xu, L., Mathiyazhagan, K., Govindan, K., Noorul Haq, A., Ramachandran, N. V., & Ashokkumar, A. (2013). Multiple comparative studies of green supply chain management: Pressures analysis. *Resources, Conservation and Recycling, 78*, 26–35. https://doi.org/10.1016/j.resconrec.2013.05.005

Yang, C., Lu, C., Jing, J., & Bernard, P. (2013). The effect of green supply chain management on green performance and firm competitiveness in the context of container shipping in Taiwan. *Transportation Research Part E, 55*, 55–73. https://doi.org/10.1016/j.tre.2013.03.005

Zailani, S., Govindan, K., Iranmanesh, M., Shaharudin, M. R., & Sia Chong, Y. (2015). Green innovation adoption in automotive supply chain: the Malaysian case. *Journal of Cleaner Production, 108*, 1115–1122. https://doi.org/10.1016/j.jclepro.2015.06.039.

Zaman, K., & Shamsuddin, S. (2017). Green logistics and national scale economic indicators: Evidence from a panel of selected European countries. *Journal of Cleaner Production, 143*, 51–63. https://doi.org/10.1016/j.jclepro.2016.12.150

Zhang, Y., Golpîra, H., & Khan, S. A. R. (2018a). The impact of GSCM on manufacturing enterprise's performance. *Journal of Advanced Manufacturing Systems, 17*(4), 445–459.

Zhang, Y., Golpîra, H., & Khan, S. A. R. (2018b). The relationship between green supply chain performance, energy demand, economic growth, and environmental sustainability: Empirical evidence from developed countries. *LogForum Scientific Journal of Logistics, 14*(4), 479–494. https://doi.org/10.17270/J.LOG.2018.304

Zhu, Q., & Sarkis, J. (2004). Relationships between operational practices and performance among early adopters of green supply chain management practices in Chinese manufacturing enterprises. *Journal of Operations Management*, 265–289.

Zhu, Q., Sarkis, J., & Lai, K. (2007). Green supply chain management: Pressures, practices, and performance within the Chinese automobile industry. *Journal of Cleaner Production, 15*(11), 1041–1052. https://doi.org/10.1016/j.jclepro.2006.05.021

Zhu, Q., Sarkis, J., & Lai, K.-H. (2008a). Confirmation of a measurement model for green supply chain management practices implementation. *International Journal of Production Economics, 111*(2), 261–273.

Chapter 3
Theoretical Framework and Methodology of GSCM

3.1 Summary of Research Objectives

This research will be conducted in the manufacturing firms of Pakistan to analyze their mechanism and find the drivers and barriers of green supply chain implementation. It will use a mixed approach. From a quantitative perspective, this research will be covered through conducting SEM (structural equation modeling) with other crucial statistical tests, and from the qualitative perspective, it will also be covered through in-depth interviews with supply chain professionals in the manufacturing industry of Pakistan. According to my best knowledge, it will be the first research which explores the barriers, drivers, and mechanism of green supply chain implementation in the perspective of the Pakistani manufacturing industry. Meanwhile, this research paper will build a new model and questionnaire based on pilot study and surveys, and all the independent variables and dependent variables will support and lead to the objective of this study. Besides, as per our expectation, some of the exogenous variables will be negatively correlated with endogenous variables of the manufacturing industry. The competitive edge can be promoted through the implementation of green supply chain sustainability, and firms can increase their profit through several exogenous variables including green manufacturing processes, ecological product design, green transportation and distribution, buying greener materials, and providing green training to their employees. Furthermore, firms can also boost their promotional activities by using green labels and also can use the sympathy of customers with green products. Finally, the implementation of green supply chain practices will not give benefits in the short run. However, it will be a long-lasting effect on firm financial stability, brand image, promotional activities, market share/market size, and government support to produce greener products.

S. A. R. Khan, *The Critical Success Factors of Green Supply Chain Management in Emerging Economies*, EAI/Springer Innovations in Communication and Computing,
https://doi.org/10.1007/978-3-030-42742-9_3

3.1.1 Theoretical Framework to Support Research Objective

Green supply chain management (GSCM) involves environmental concerns for traditional supply chain management (Gilbert, 2000). The definition of GSCM in published papers has ranged from GP (green purchasing) to integrated GSCM from supplier to producer to customer (Zhu & Sarkis, 2004). Many research papers have examined the GSCM within different components including ED (ecological design), GM (green manufacturing), GP (green purchasing), GT (green transportation), GD (green distribution), and a mixture of these components (Khan & Dong, 2017; Khan, Dong, & Zhang, 2018; Khan et al., 2019; Khan & Zhang, 2019). As an equation, GSCM contains GP, GM, GD, and RL (Hervani et al., 2005).

As an excellent starting point for discussion, institutional theory can be used to explain the significant drivers for implementing GSCM practices (DiMaggio & Powell, 1983). The institutional theory states that companies adopt initiatives in order to achieve legitimacy and social acceptance. Accordingly, normative, mimetic, and coercive pressures can be identified as external drivers that create pressure on firms to adopt GSCM practices.

- The normative pressures exerted by companies on similar people or industries in similar groups. For the adoption of the GSCM practices, the information on costs and benefits of implementing green practices are likely to be shared within the same industry (Colicchia, Creazza, & Dallari, 2017). Sharing these norms among members of a similar network facilitates potential influence on implementing GSCM practices. Therefore the prevalence of GSCM practices within the industry is used as a vital driver of normative pressure.
- The coercive pressures exerted by external bodies through informal and formal power. The regulatory authorities, government laws, and regulations are the best examples of normative pressures that influence organizational behavior. The published literature supports the regulatory forces as coercive pressure (Bakeshlou, Khamseh, Asl, Sadeghi, & Abbaszadeh, 2017; DiMaggio & Powell, 1983). Thus sustainability and other laws and regulations are used as the main driver for coercive pressure.
- Mimetic pressures arise when a firm follows successful competitors in the industry. Companies may imitate successful competitors through benchmarking in terms of reducing the risk of implementing GSCM practices and achieve competitive advantage/second-mover advantages (Basiri & Heydari, 2017). Therefore perceived success of GSCM is used as the main driver for mimetic pressures. Although there are sufficient researches based on institutional theory, many different types of research have used different drivers for implementing GSCM practices (Golini, Moretto, Caniato, Caridi, & Kalchschmidt, 2017). The drivers are defined as stimuli that encourage firms to implement GSCM practices. The six critical drivers of GSC strategy adoption are based on the institutional theory (Golini et al., 2017). These are perceived success of green supply chain pioneers, regulatory factors, the extent of strategic alignment with customers who implemented green supply chain strategy, the extent of strategic (Ghosh, Jha, &

Sarmah, 2017) alignment with suppliers or vendors who implemented green supply chain strategy, prevalence of green supply chain strategy implementation within the industry, and public concerns (Fallahpour, Udoncy Olugu, Nurmaya Musa, Yew Wong, & Noori, 2017).

- The research conducted by Khidir and Zailani (2011) has identified 30 research papers that empirically investigated the drivers for green practices as customer pressure, laws and regulation, expected business benefits, social responsibility, community pressure, competition, and employee pressure (Herold & Lee, 2017). On the other hand, the findings of the research paper uncover that laws and regulations significantly encourage the implementation of the GSCM practices, expected business (Hoekstra, 2017), social responsibility, and customer pressure. Even though this empirical study has identified that there is an apparent effect of the above key four drivers, Bowen, Cousins, Lamming, and Farukt (2001) have explored that there is no clear association between green purchasing and laws/regulations. Zhu and Sarkis (2004) conducted the study which found no (Habidin et al., 2017) such significant relationships between regulatory, supplier pressures, and market and green supply chain initiatives.

As a support for identifying internal motivating factors for implementing GSCM practices, the *resource-based theory* has been used (Macchion et al., 2017). This theory asserts that green competitive edge can be gained through harnessing the resources which are rare, valuable, imperfectly imitable, and non-substitutable (Loosemore & Lim, 2017). The resources, in turn, contain three subgroups, including intangible resources, competencies, and tangible resources. The tangible resources included current and fixed assets that have long-run capacity (Miroshnychenko, Barontini, & Testa, 2017), for example, equipment, land, plant, and inventory stock. The intangible resources cover intellectual property such as patents and trademarks, reputation, brand image, and secure networks (Habidin et al., 2017). The competencies encompass skills of individuals or groups as well as companies' routines and (Rueda, Garret, Lambin, Garrett, & Lambin, 2017) interactions through which all the companies' resource coordinated (Grant, 1991).

Besides, knowledge and capabilities for the adoption of GSCM practices of a company are considered as resources according to the resource-based view (Lai et al., 2010; Qin, Liu, & Pedrycz, 2017). Another influential internal factor is senior management commitment towards the implementation of green supply chain practices. The senior management is primarily critical for GSCM practices since it has an extraordinary ability to influence and support for adopting GSCM practices (Rezaei & Kheirkhah, 2017; Yang & Xiao, 2017). Walker, Di Sisto, and McBain (2008) and Vinyes et al. (2017) recognize economic benefit as an internally motivating factor for adopting GSCM practices and identified reputation and brand image as an economic benefit achieved through GSCM practices (Song, Ming, & Liu, 2017). Therefore, the author considered firm reputation and brand image as a significant critical internal driver in the resource-based view (Shaharudin, Govindan, Zailani, Tan, & Iranmanesh, 2017).

The theory of reverse logistics is also called four Rs: reduce, reuse, recycle, and recover energy and disposal in landfills.

The flow of reverse products requires proper forecasting as well as place/point for collecting the products for reprocessing and disposing of products/components on the return trip. In the hierarchy of reverse logistics, the very responsible option in the first place is to reduce resources in use. As far as they are into their business, companies can incorporate principle in the following ways:

- Design the network of logistics (warehouse and transport), which requires less energy and resources (it is cost control directly).
- Design the products as energy-efficient in use as well as during manufacturing, which requires less consumption of energy.
- Companies can design products that are more efficient in the usage of physical resources.

The landfill is the least desirable option. However, there are some physical products which must go to landfill (Fiorini & Jabbour, 2017). But before landfill construction, first, select the best available land for this purpose. The reverse journey of products should (Zhang & Wang, 2017) be responsible for landfill. Because off it prevents degrading items leaching from water pollution or environmental pollution (Blome, Foerstl, & Schleper, 2017). Furthermore, the stakeholder theory also supports the green supply chain concept, for which the theory states that stakeholders are the individuals and groups who can be affected by a firm's actions or a firm's performance (Blome et al., 2017). Stakeholder theory suggests that a firm produces externalizations, which can cause stakeholders to put pressure on firms to reduce adverse effects and thus influence organizational practice (Thöni & Tjoa, 2015; Yenipazarli, 2016).

The theory of stakeholder is a theory of business ethics and organizational management that addresses values and morals in managing a company. It was discussed by Ian Mitroff in his book *Stakeholders of the Organizational Mind*, which was published in 1983 in San Francisco. *The Freeman* had a research paper on *stakeholder theory* in the *California Management Review* in late 1983.

In the traditional view of a firm, the theory of stakeholder view, only the shareholders or owners are essential, and the (Yenipazarli, 2016) firm has a binding fiduciary duty to put their needs first (Diaz & Marsillac, 2016) to increase the value for them. The theory of stakeholder instead argues that there are other members involved, including customers (Kumar, Rahman, & Chan, 2016), financers, governmental and legal authorities (Braithwaite, 2012), suppliers, employees, trade unions, political groups, and trade associations. Even competitors are sometimes counted as stakeholders, too – their status being (Leclerc & Partners, 2013) derived from their capacity to affect the company and its stakeholders.

The theory of stakeholder view of strategy integrates both a market-based view (Xia et al., 2017); (Fichtinger, Ries, Grosse, & Baker, 2015) and a resource-based view and adds a social-political level. One prevalent version of stakeholder theory seeks to define the specific (Xia et al., 2017) stakeholders of a firm (the normative theory of stakeholder identification) and then examine (Rezaei, Nispeling, Sarkis, &

Tavasszy, 2016) the conditions under which senior management treats these members as stakeholders (the descriptive theory of stakeholder salience).

A typology derives of stakeholders based on (Cucchiella, D'Adamo, & Lenny Koh, 2015) the attributes of power (the extent a member (Mraihi & Abid, 2013) as a means to impose its will in a relationship), legitimacy (Simão, Gonçalves, & Taboada Rodriguez, 2016) (expected structures and socially accepted or behaviors), and urgency (criticality of the stakeholder's claims or time-sensitivity) (Fiore, Silvestri, Contò, & Pellegrini, 2016). By examining the combination of these attributes in a binary manner, eight types of stakeholders are (Fahimnia, Sarkis, Choudhary, & Eshragh, 2015) derived along with their implications for the firm. Friedman and Miles (Tseng, Lin, Tan, Chen, & Chen, 2014) identify the implications of contentious relationships between organizations and stakeholders by introducing conflicting or compatible interests and contingent/necessary connections (Meixell & Luoma, 2015) as additional attributes with which to explore the configuration of these relationships. Robert Allen Phillips distinguishes between derivatively (Seles, de Sousa Jabbour, Jabbour, & Dangelico, 2016) legitimate stakeholders (those whose stakeholder status is derived from their ability to affect the firm or its normatively legitimate stakeholders) (Seles et al., 2016) and normatively legitimate stakeholders (those to whom a firm holds a moral obligation).

The theory of stakeholder is prevalent in the business ethics, green supply chain, and corporate social responsibility field. For example, the GRI (Global Reporting Initiative) and ISO 26000 cover stakeholder analysis.

In a field such as management, law, and human resource, stakeholder theory is also very successful in challenging the common analysis frameworks by suggesting to put stakeholders' needs at the starting point of any action. Several researchers, such as Geoffroy Murat, tried to apply stakeholder's theory to irregular welfare.

The leading theory of the green supply chain is to protect environmental sustainability and increase awareness of the clean environment. GSCM practices are essential for the success of this century's businesses whenever we discuss green supply chain management, so there are some fundamental factors that play a significant role, such as consumer pressure, governmental regulatory authority, and international community pressure.

In business, supplier sell materials to manufacturers, and after production goods/products go to DC (distribution center), and then those goods are sold to the customer/consumers and after undergo PLC (product life cycle); there are mainly three options whether products go to the recycling process, landfill, or incineration plant (Zhao, Liu, Zhang, & Huang, 2017).

The operations of the traditional supply chain consume much electricity, and due to the higher consumption of electricity, it also emits higher emissions in the environment through polluted supply chain networks. Many developed countries, including Germany, the USA, and the UK, have started consuming renewable energy and biofuel electricity to reduce harmful effects on environmental sustainability. Fig. 3.1 indicates the sources of energy and their contribution to global consumption. Eighty-four percent of energy is produced by nonrenewable sources including nuclear energy, fossil fuel energy, coal, oil, and natural gas energy. On the

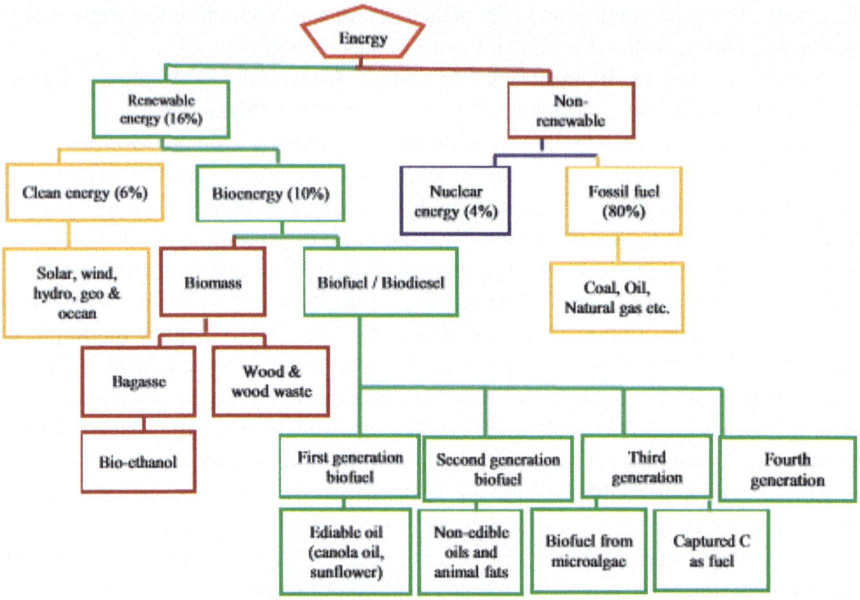

Fig. 3.1 Energy sources and global consumption

other hand, 16% of electricity is produced by renewable energy sources mainly including clean energy and bioenergy. Several developed countries such as Germany, the USA, and Japan have shifted towards renewable energy sources, while many countries, mainly in the Asian region including Pakistan, India, and China, are shifting towards renewable energy.

Many researchers claimed that renewable energy sources and green practices are directly improving the country's economic activities and foreign direct investment inflows. On the other hand, sustainable supply chains not only process protection for environmental sustainability and reduce substantial cost but also increase tourism opportunities (AlKhidir & Zailani, 2009; Khan et al., 2017). The word green practices is used in a vast meaning, and green practices covered mainly functions of the green supply chain, including procurement, logistics, distribution, selling, and manufacturing. Many times the words green supply chain and green logistics are used interchangeably. However, the main cracks of both words are the same to protect environmental sustainability. In Fig. 3.2, we can see that green logistics components are directly affecting economic efficiency, competitiveness choice, environmental externalization, climate change, and carbon emissions.

The sustainability supply chain process cannot be completed without the involvement of green practices, while the risk factor is also involved in this end-to-end process. However, firms can reduce the risk through better planning expertise and provide training to the employees. In the whole green supply chain process, the top management role is very critical, and the commitment of senior leadership can complete the green projects.

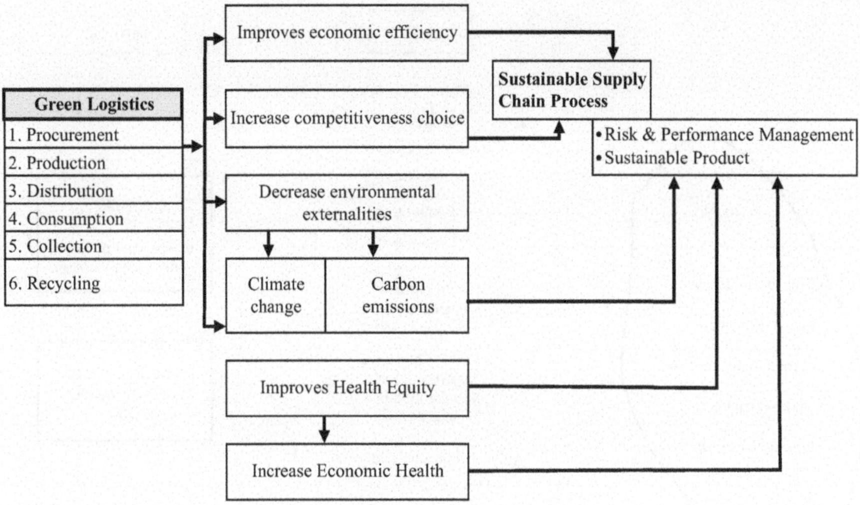

Fig. 3.2 Green practices and environmental sustainability

The regulatory policy and law enforcement with customs regulation have a healthy relationship; market sophistication has strong relationship with international shipment, logistics quality and competence, and tracking and tracing systems of logistics; human capital association with logistics quality and competence and tracking and tracing of logistics systems are also significantly critical; the economy of the country and/or industries of the country has a significant relationship with the infrastructure of logistics, and logistics goods arrived on time. On the other hand, infrastructure and environmental health and ecosystems have a significant association, while timeliness with environmental health and ecosystems also have a significant relationship. In Fig. 3.3, we can see the complete picture of the green practices and laws and regulations relationships with logistics and environmental health growth.

The environmental supply chain is connected with several echelons in the society, namely, raw material supplier, component suppliers, transport, manufacturer, distributor/wholesaler, retailer, consumer, incinerators, and landfill disposal. Fig. 3.4 shows the complete mechanism of the manufacturing supply chain. On the other hand, it also shows the steps of the manufacturing supply chain.

Whenever the lead-time is long or, raw materials buy from foreign suppliers so that it will direct emits more Air emissions, consume higher fossil fuel, and destroy the environmental sustainability with flora and fauna lives.

There are several factors, mostly external factors and some internal factors, which contribute to the firms' improved performance. External factors included government support, regulatory authority laws and regulations, consumer pressure and demand, and international market demand. On the other hand, internal factors included organizations inside, such as processes, management seriousness, and employee training. Technological factors are usually called external factors, but

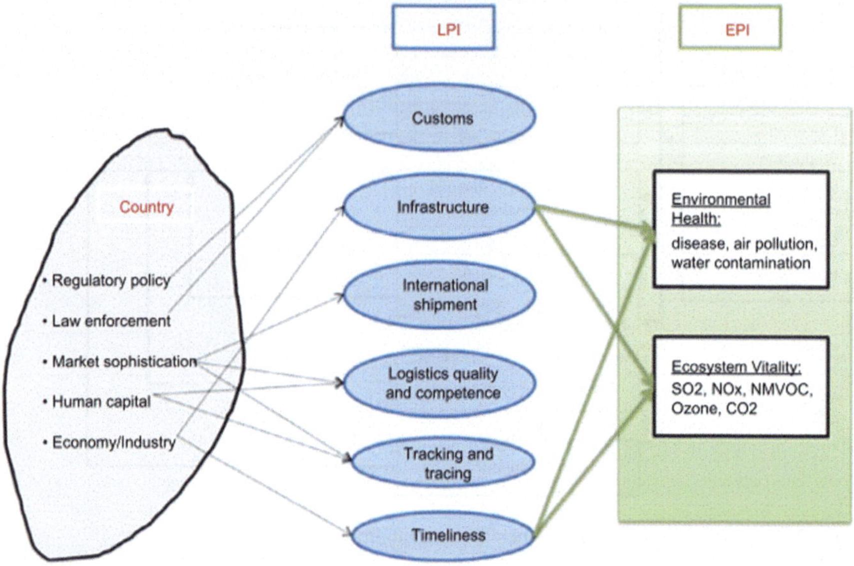

Fig. 3.3 Environmental performance and supply chain performance

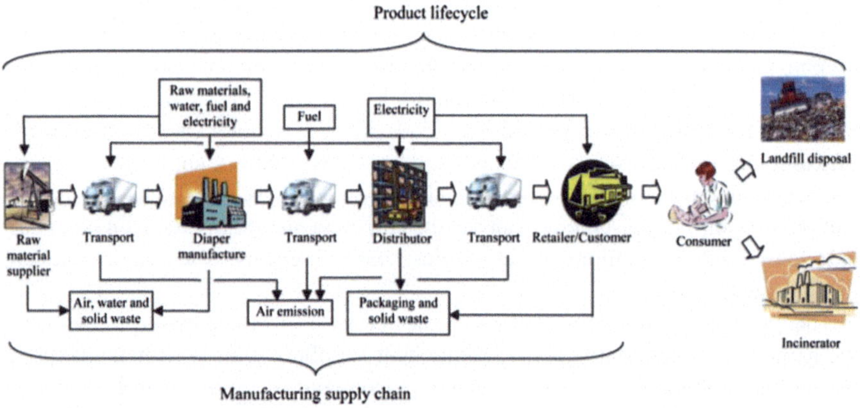

Fig. 3.4 Manufacturing supply chain mechanism

somehow, it can also be count as an internal factor if it is customized products. Fig. 3.5 shows that the main contribution to the success of the firms is related to external environmental factors and organizational and technological factors. In short, firms cannot ignore the customer requirement and pressure, regulatory authorities' policies, senior management commitment, and employee training which are playing a vital role in the improvement of the firms' performance.

The GSCM (green supply chain management) is a complete concept of green environmental sustainability to protect fauna and flora lives. In the GSCM process,

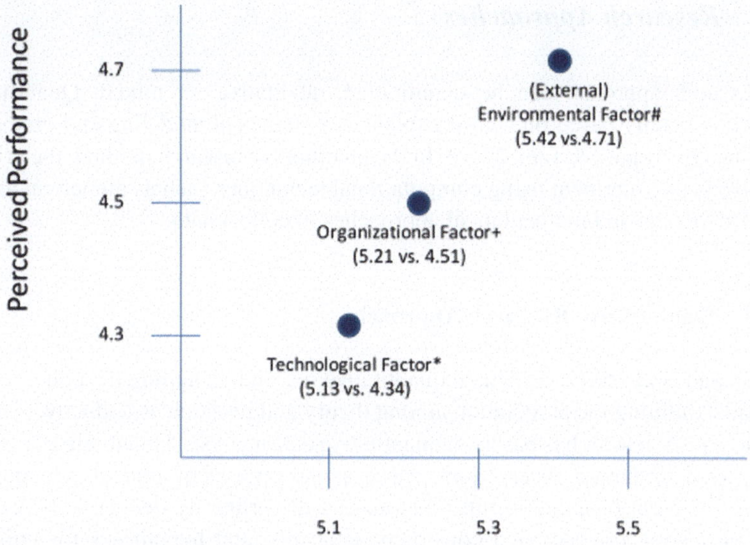

Fig. 3.5 Contribution factors to performance

the following are the critical processes involved: green design, green purchasing, green transportation, green logistics, and reverse logistics. Furthermore, in detail, we can see that green design included regulations, environmental performances, and economic performance; green purchasing included supplier-customer collaboration, enforcement of stakeholders, and quality of regulations; green transformation included green manufacturing, green packaging, and green stock politics; green logistics covers organization of logistics networks, quality of service, and quality of technology; and reverse logistics included reducing activities, recycling, re-manufacturing, reusing, and disposal.

3.2 Research Design

To obtain the research problems, the researcher adopts a plan of investigations to the research questions. This research is based on phenomenological design. This study attempts to understand respondents' perspectives, perceptions, and specific phenomenon. In short, in phenomenological design, researcher's objective is to understand experiences and nature. If researchers want to find the relationship between variables, then the researcher should use co-relational design. However, in this research, we are going to use both phenomenological and co-relational designs.

3.2.1 *Research Approaches*

The research approach can be quantitative, qualitative, or mixed. Quantitative research is mainly based on causal-explanatory statistical modeling and predictive modeling (Evermann & Tate, 2016). In the quantitative research method, the critical objective is measurement using computational techniques such as numerical analysis and statistical and mathematical approaches to collect data.

3.2.1.1 Quantitative Research Approach

Bryman and Bell (2005) explained quantitative research, entailing the collection of data and exhibiting the association among theory and research as deductive, so this specific type of research is based on quantitative data analysis. Quantitative research mainly used structured observations, surveys, and reviews of authentic documents for the collection of numeric information, and the primarily deductive process is based on testing pre-specified constructs, concepts, and hypotheses for building theory (Cepeda Carrin, Henseler, Ringle, & Roldn, 2016). Research is more on the breadth and less on in-depth information across several observations; data is on numeric forms, analysis, and statistical tests that are used for testing or measuring hypotheses to achieve the research objective.

(Cepeda Carrin et al., 2016) Quantitative research is mainly used to test the hypothesis developed based on literature reviews and to quantify opinions, attitudes, and behaviors and generalize findings from a larger sample of the population. Moreover, quantitative research is based on measurable data to formulate facts; methods are very structured than qualitative research. Usually, data are collected through paper surveys, online surveys, kiosk surveys, mobile surveys, structured interviews, etc. (Richter, Cepeda, Roldán, & Ringle, 2016).

3.2.1.2 Qualitative Research Approach

Qualitative research approach is primarily exploratory. It is used to achieve an in-depth understanding of underlying opinions, phenomena, and motivations. It also offers insights into the issues or support to build hypotheses and ideas for potential quantitative research; moreover, it is also used to uncover trends in thoughts and dive deeper into the research issues/problems/questions. The data collection methods used in qualitative research is mainly based on semi-structured and/or unstructured interviews and individual interviews (Nitzl & Roldán, 2016). The size of the research sample is usually smaller, and respondents are nominated to attain a given requirement.

In qualitative research, primary techniques are an inductive process to formulate hypothesis or theory and to describe a condition and/or problem, in-depth information on smaller observations, and semi-structured and unstructured response options. Mainly, data is in text form, and no such statistical or mathematical tests are used. Largely research depends on rigor, skills, and knowledge of the researchers. Finally, the findings of the research are specifically implacable and/or less generalized (Nitzl, 2016).

3.2.1.3 Mixed Research Approach

Mix approach in the research, this research approach that how we conduct it is a mixed approach, and in the mixed approach, we use statistical analysis techniques and tools as well as describe situations and phenomena during in-depth interviews with supply chain personnel. Mixed methodology involves analyzing, collecting, and integrating qualitative and quantitative (text-based and interviews (unstructured and/or semi-unstructured)) research (Evermann & Tate, 2016). In the mixed approach of research, qualitative and quantitative research techniques are merged to enhance research reliability and achieve the maximum research objective and for the researcher to attain in-depth and in breadth corroboration, understanding while offsetting the weaknesses inherent to using single research approach by itself. The key advantage of mix research approach is the possibility of triangulation. Triangulation allows us to recognize aspects of a phenomenon more in-depth and accurately by approaching it from different vantage points using different techniques (Weerawardena, Mort, Salunke, Knight, & Liesch, 2014). Successful triangulation needs in-depth and careful analysis of the type of information offered by each research approach, including its weaknesses and strengths.

In short, mix research approach used both techniques, quantitative and qualitative, to increase the reliability of the research and findings. This research study is based on mix approach, so the researcher will find and explain the motivating drivers and barriers of green environmental supply chain practices from the perspective of Pakistani manufacturing firms.

3.2.2 Research Methods

In this research, the researcher will cover the exploratory and explanatory part. As far as green supply chain management is concerned, the researcher wants to explain and explore the motivating drivers/factors and barriers which firms face in the execution of green supply chain management practices in manufacturing firms of Pakistan.

For data collection, the researcher adopted/selected the survey method and in-depth interviews with senior managers and professionals of supply chain management. The source of data can be primary and secondary. However, in this research study, the researcher has used the primary data source to enhance the reliability and validity of data to achieve the research objective and find the innovations points. Furthermore, the researcher is going to use a questionnaire (Likert scale), and in-depth interviews will also be conducted in the pilot phase of the research study.

3.2.3 Research Context

The research purpose means on behalf of which purpose we are going to conduct a research study. Research purposes can be descriptive-based, explanatory, and exploratory, i.e., how people come to understand the process of researching a concept, phenomena, or idea (Richter et al., 2016). There are three main approaches: exploratory, explanatory, and descriptive researches (Richter et al., 2016).

In this research, the researcher will cover exploratory and explanatory purposes as far as the researcher wants to explore and explain the motivating drivers/factors and barriers which organizations face in the implementation of environmentally friendly practices from the perspective of Pakistani manufacturing firms.

3.2.4 Research Process

A data collection instrument is used for the collection of data. The researcher must ensure that the questionnaire/instrument is reliable and valid. The reliability and validity of any research study are mainly based on a large extent the suitability of the questionnaire/instrument of the research study. For data collection, the researcher has selected the survey method and in-depth interviews with senior managers and professionals of supply chain management. In this research study, the researcher is going to use a questionnaire (Likert scale), and in-depth interviews will also be conducted with senior supply chain managers and professionals to fulfill the objective of this study. The questionnaire used the 7-point scale: 7, strongly agree, and 1, strongly disagree. This study will use mix research approach in which we used statistical analysis techniques and tools as well as described phenomena and situations during in-depth interviews with supply chain managers and professionals.

3.3 Discussion of Current Research Methods

This questionnaire is used to collect data from manufacturing firms. The questionnaire is attached in Appendix B.

3.3.1 *Questionnaire Survey*

We have conducted interviews with senior managers and directors of the firms for the qualitative part of this study and unstructured interviews to find barriers and drivers of green supply chain practices in the context of manufacturing firms of Pakistan. The unstructured interview questions are attached in Appendix C.

3.3.2 *Unstructured Interviews*

The interviews are conducted with the supply chain directors and managers, who mostly have more than 10–20 years of work experience in Pakistani manufacturing firms. It was intended to find the barriers and difficulties faced during the implementation of green supply chain practices or the problems faced during converting traditional supply chain into green/environmental-friendly supply chain. After the in-depth interviews, we prepared "after the interview script" for analysis purposes (see Table 3.1 for the respondents' profile).

3.4 Choice of Research Locations and Interviewees

The researcher has conducted a study in Pakistan and collected the samples from different manufacturing firms for testing hypotheses and fulfilling the objectives of this research. Mainly, the researcher collected the data through a questionnaire survey for the quantitative part of the research, while for the qualitative part of the research, the researcher conducted in-depth interviews with supply chain directors, supply chain managers, and supply chain supervisors.

3.5 Statistical Techniques

In a research study, usually, researchers used some statistical techniques and/or methods to achieve the objective of quantitative research. In statistics, we differentiate the most applied methods among the existing ones: design of experiments, descriptive statistics, sampling, hypothesis building, regression, correlation, factor analysis, structural equation modeling, time series analysis of the measurement accuracy, KMO tests, t-test, ANOVA test, and confirmatory and exploratory factor analyses.

Table 3.1 Respondents' profile

Respondents' profile	Number of respondents	%
Job title		
Supply chain professionals	41	50.6
Supply chain managers	29	35.8
Supply chain directors	11	13.6
Gender		
Male	49	49.4
Female	32	39.5
Respondents' age		
Less than 25	14	17.3
26–35	31	38.3
36–45	17	21.0
46–55	11	13.6
56 and above	8	9.9
Job experience		
Less than 3 years	7	8.6
3–10 years	26	32.1
11–15 years	21	25.9
16–20 years	14	17.3
More than 20 years	13	16.0
Number of employees		
Less than 50 workers	43	18.0
50–100 workers	125	55.0
101–200 workers	138	15.0
201–500 workers	23	8.0
More than 500 workers	16	5.0

In this research study, we are going to adopt SEM (structural equation modeling). In general structural equation model, the term used to describe a vast statistical (numbers) models and evaluation, check the validity of fundamental theories. The researcher will use structural equation model as a statistical tool for the quantitative part of this study.

3.6 Sampling Techniques

The researcher is going to use purposive sampling for this research. Purposive sampling is also called purposeful sampling; it is like the convenience. Random sampling and procedure is the same as probability sampling where the researcher collects a sample from significant populations. In the first stage, the researcher identified the subdivision and subgroups of the population on behalf of interest, and then from each subdivision, the researcher selects cases in a purposive manner.

3.7 Data Source

3.7.1 *Questionnaire Analysis*

Questionnaires are distributed to the respondents after a brief introduction of the purpose of the study. In the first round, 220 forms were distributed in different manufacturing firms, and filled forms received were 186. The second round was conducted during summer vacations, and 280 forms were distributed again in the different manufacturing firms, and also reminders were sent to the first round respondents. Finally, the complete 500 forms were distributed in the Pakistani manufacturing firms during summer and winter vacations, and the total filled forms received from the respondents were 415, and the total response rate was 83%. So finally, 415 filled forms have been included in the dataset for further analysis purposes. This number of responses is satisfactory for data analysis because the vital purpose of this study has been to validate a proposed model, not to verify an existing model (Oh, 2000; Huang, Schrank & Dubinsky, 2004).

3.7.2 *Primary Data*

The data source of a research paper can be primary or secondary. The primary data is information which researcher collects specifically for the research purpose. The advantages of this data are specific and customized and based on the requirement of research. On the other hand, primary data can be expensive to obtain in terms of money and time because most of the time, researchers need several months or weeks to collect data from firms or individuals. In this research study, primary data has been used, which is collected by researchers through in-depth interviews (unstructured) with supply chain professionals and managers.

3.7.3 *Secondary Data*

Secondary data are collected by someone such as governments/regulatory authority, NGOs (non-government organizations), stock exchange, banks, financial institutions, censuses, firms' records, or any other firm/person. The advantage of this data is inexpensive in terms of money and time as compared to primary data since secondary data can be downloaded from international NGOs, stock exchange records, firms' published reports, the World Bank, IMF (International Monetary Funds), UNESCO, and UN (United Nations).

3.8 Research Plan

The research plan means is followed by a researcher to fulfill the requirement and objective of the research study. This research plan has been explained in chart form (see Fig. 3.6).

3.9 Research Model

This research study is based on structural equation modeling for the exploratory part, and the remaining is based on unstructured interview part. The researcher develops models and hypotheses after extensive literature reviews and discussions with industry experts. However, a pilot study was conducted before for checking the research instrument, and after an improvement full flash survey was conducted for collecting data to achieve the research objective (see Fig. 3.7).

3.10 Research Hypotheses

H-1: Internal environmental management significantly and positively improved green information systems.

H-2: IEM (internal environmental management) has a significant and positive association with GM (green manufacturing).

H-3: IEM (internal environmental management) has a significant and positive effect on GD (green distribution).

H-4: IEM (internal environmental management) has a significant and positive correlation with GT (green transportation).

H-5: IEM (internal environmental management) has a significant and positive effect on product ED (ecological design).

H-6: GIS (green information systems) have a significant and positive impact on GM (green manufacturing).

H-7: GIS (green information system) has a significant and positive effect on product GD (green distribution).

H-8: GIS (green information systems) have a significant and positive impact on GT (green transportation).

H-9: GIS (green information system) has a significant and positive effect on product ED (ecological design of products).

H-10: GM (green manufacturing) positively and significantly enhanced ECP (economic performance of the firms).

H-11: GM (green manufacturing) has a significant and positive impact on firms' CA (competitive edge of the firms).

H-12: GM (green manufacturing) positively and significantly enhanced ENP (environmental performance of the firms).

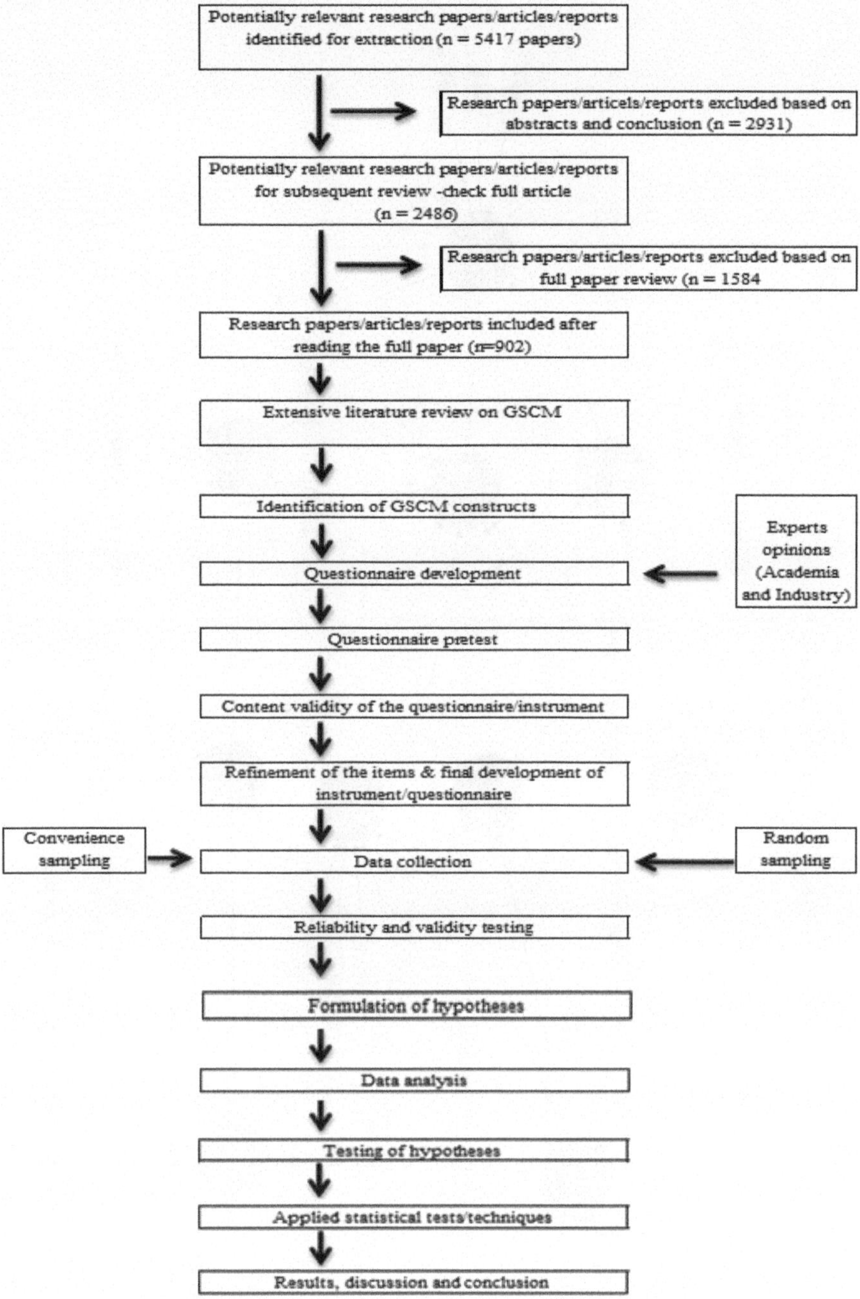

Fig. 3.6 Research plan

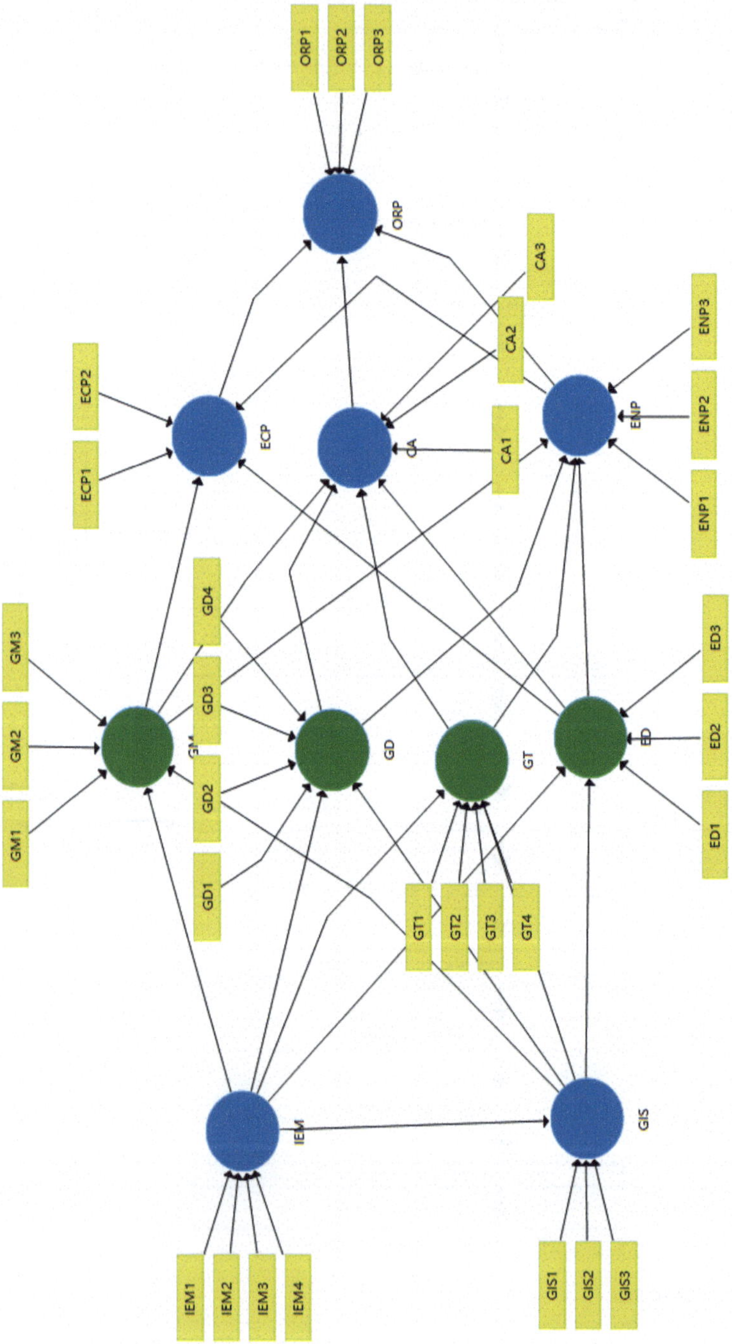

Fig. 3.7 Research measurement model for hypotheses testing

H-13: GD (green distribution) has a significant and positive effect on firms' CA (competitive edge).

H-14: GD (green distribution) positively and significantly improved ENP (environmental performance of the firms).

H-15: GT (green transportation) has a significant and positive effect on firms' CA (competitive edge).

H-16: GT (green transportation) positively and significantly improved ENP (environmental performance of the firms).

H-17: ED (ecological design) has a significant and positive effect on firms' ECP (economic performance).

H-18: ED (ecological product design) positively and significantly created CA (competitive edge for the companies).

H-19: ED (ecological product design) positively and significantly improved the environmental performance of the companies.

H-20: ENP (environmental performance) is positively and significantly enhanced ECP (economic performance).

H-21: ENP (environmental performance) is positively and significantly enhanced ORP (organizational performance).

H-22: CA (competitive advantage) is positively and significantly enhanced ORP (organizational performance of the manufacturing firms).

H-23: ECP (economic performance) is positively and significantly enhanced ORP (organizational performance of the manufacturing firms).

3.11 Overview of Research Methodology

The research will find and explain the motivating drives and barriers/hurdles of green supply chain management implementation from the perspective of Pakistani manufacturing firms. In this research, the researcher will cover the exploratory and explanatory parts. The researcher wants to explain and explore the motivating drivers/factors and barriers which firms face in the execution of green supply chain management practices in manufacturing firms of Pakistan.

For data collection, the researcher has adopted the survey method and in-depth interview method with senior managers and professionals of supply chain management. The source of data can be primary and secondary. Nevertheless, in this research study, the researcher has used the primary data source to enhance the reliability and validity of data to achieve the research objective and find the innovation points. Furthermore, the researcher is going to use a questionnaire (Likert scale), and in-depth interviews will also be conducted in the pilot phase of the research study. This study mainly focuses on manufacturing firms in Pakistan. It will cover all the key aspects to explore the mechanism, barriers, and drivers of green supply chain execution and the effect of green supply chain management practices on manufacturing firms' performance in the context of Pakistan.

References

AlKhidir, T., & Zailani, S. (2009). Going green in supply chain towards environmental sustainability. *Global Journal of Environmental Research, 3*(3), 246–251.

Bakeshlou, E. A., Khamseh, A. A., Asl, M. A. G., Sadeghi, J., & Abbaszadeh, M. (2017). Evaluating a green supplier selection problem using a hybrid MODM algorithm. *Journal of Intelligent Manufacturing, 28*(4), 913–927. https://doi.org/10.1007/s10845-014-1028-y

Basiri, Z., & Heydari, J. (2017). A mathematical model for green supply chain coordination with substitutable products. *Journal of Cleaner Production, 145*, 232–249. https://doi.org/10.1016/j.jclepro.2017.01.060

Blome, C., Foerstl, K., & Schleper, M. C. (2017). Antecedents of green supplier championing and greenwashing : An empirical study on leadership and ethical incentives. *Journal of Cleaner Production, 152*, 339–350. https://doi.org/10.1016/j.jclepro.2017.03.052

Bowen, Cousins, Lamming, & Farukt. (2001). The role of supply management capabilities in green supply. *Production and Operations Management, 10*(2), 174–189.

Braithwaite, A. (2012). Global logistics: new directions in supply chain management. Kogan Page Limited. https://doi.org/10.1007/s13398-014-0173-7.2

Bryman & Bell (2005). Social Research Methods, 4th edition, ISBN 978-0-19-958805-3, Oxford University Press Inc., New York.

Cepeda Carrin, G., Henseler, J., Ringle, C. M., & Roldn, J. L. (2016). Prediction-oriented modeling in business research by means of PLS path modeling: Introduction to a JBR special section. *Journal of Business Research, 69*(10), 4545–4551. https://doi.org/10.1016/j.jbusres.2016.03.048

Colicchia, C., Creazza, A., & Dallari, F. (2017). Lean and green supply chain management through intermodal transport: Insights from the fast-moving consumer goods industry. *Production Planning & Control, 28*(4), 1–14. https://doi.org/10.1080/09537287.2017.1282642

Cucchiella, F., D'Adamo, I., & Lenny Koh, S. C. (2015). Environmental and economic analysis of building integrated photovoltaic systems in Italian regions. *Journal of Cleaner Production, 98*, 241–252. https://doi.org/10.1016/j.jclepro.2013.10.043

Diaz, R., & Marsillac, E. (2016). Evaluating strategic remanufacturing supply chain decisions. *International Journal of Production Research, 7543*(October), 1–18. https://doi.org/10.1080/00207543.2016.1239848

DiMaggio, P. J., & Powell, W. (1983). The iron cage revisited institutional isomorphism and collective rationality in organizational fields. *American Sociological Review, 48*, 147–60.

Evermann, J., & Tate, M. (2016). Assessing the predictive performance of the structural equation model estimators. *Journal of Business Research, 69*(10), 4565–4582. https://doi.org/10.1016/j.jbusres.2016.03.050

Fahimnia, B., Sarkis, J., Choudhary, A., & Eshragh, A. (2015). Tactical supply chain planning under a carbon tax policy scheme: A case study. *International Journal of Production Economics, 164*, 206–215. https://doi.org/10.1016/j.ijpe.2014.12.015

Fallahpour, A., Udoncy Olugu, E., Nurmaya Musa, S., Yew Wong, K., & Noori, S. (2017). A decision support model for sustainable supplier selection in sustainable supply chain management. *Computers and Industrial Engineering, 105*, 391–410. https://doi.org/10.1016/j.cie.2017.01.005

Fichtinger, J., Ries, J. M., Grosse, E. H., & Baker, P. (2015). Assessing the environmental impact of integrated inventory and warehouse management. *International Journal of Production Economics, 170*, 717–729. https://doi.org/10.1016/j.ijpe.2015.06.025

Fiore, M., Silvestri, R., Contò, F., & Pellegrini, G. (2016). Understanding the relationship between green approach and marketing innovations tools in the wine sector. *Journal of Cleaner Production, 142*, 4085–4091. https://doi.org/10.1016/j.jclepro.2016.10.026

Fiorini, P., & Jabbour, C. (2017). Information systems and sustainable supply chain management towards a more sustainable society: Where we are and where we are going. *International Journal of Information, 37*(4), 241–249. https://doi.org/10.1016/j.ijinfomgt.2016.12.004

Ghosh, A., Jha, J. K., & Sarmah, S. P. (2017). Optimal lot-sizing under strict carbon cap policy considering stochastic demand. *Applied Mathematical Modelling, 44*, 688–704. Elsevier Inc.. https://doi.org/10.1016/j.apm.2017.02.037

Gilbert, N. (2000). Models, Processes and Algorithms: Towards A Simulation Toolkit. In: Suleiman R., Troitzsch K. G., & Gilbert N. (eds) Tools and Techniques for Social Science Simulation. Physica, Heidelberg.

Golini, R., Moretto, A., Caniato, F., Caridi, M., & Kalchschmidt, M. (2017). Developing sustainability in the Italian meat supply chain: An empirical investigation. *International Journal of Production Research, 55*(4), 1183–1209. https://doi.org/10.1080/00207543.2016.1234724

Grant, R. M. (1991). The resource-based theory of competitive advantage: implications for strategy formulation. *California Management Review, 33*(3), 114–135.

Habidin, et al. (2017). Lean manufacturing practices, ISO 14001, and environmental performance in Malaysian automotive suppliers. *International Journal of Management Science and Engineering Management, 13*(1), 45–53.

Herold, D. M., & Lee, K.-H. (2017). Carbon management in the logistics and transportation sector: An overview and new research directions. *Carbon Management, 8*, 79–97. https://doi.org/10.1080/17583004.2017.1283923

Hervani, A., Helms, M., & Sarkis, J. (2005). Performance measurement for green supply chain management. *Benchmarking: An International Journal, 12*(4), 330–353.

Hoekstra, A. Y. (2017). Water footprint assessment: Evolvement of a new research field. *Water Resources Management, 31*, 3061–3081. https://doi.org/10.1007/s11269-017-1618-5

Huang, Schrank, & Dubinsky. (2004). Effect of brand name on consumers' risk perceptions of online shopping. *Journal of Consumer Behaviour, 4*(1), 40–50.

Khan, S. A. R., & Dong, Q. (2017). Impact of green supply chain management practices on firms' performance: an empirical study from the perspective of Pakistan. *Environ Sci Pollut Res 24*, 16829–16844. https://doi.org/10.1007/s11356-017-9172-5.

Khan, S. A. R., Dong, Q., & Zhang, Y. (2018). The impact of green supply chain practices in business performance: Evidence from Pakistani FMCG firms. *Journal of Advanced Manufacturing Systems, 17*(2), 267–275.

Khan, S. A. R., Dong, Q., Zhang, Y., & Khan, S. S. (2017). The impact of green supply chain on enterprise performance: In the perspective of China. *Journal of Advanced Manufacturing Systems, 16*(3), 263–273.

Khan, S.A.R., Zhang, Y. (2019) Strategic supply chain management, EAI/springer innovations in communication and computing Series, Springer Publisher, ISBN 978-3-030-15058-7. https://www.springer.com/gp/book/9783030150570

Khan, S. A. R., Sharif, A., Golpira, H., & Kumar, A. (2019). A green ideology in Asian emerging economies: From environmental policy and sustainable development. *Sustainable Development, 27*(6), 1063–1075.

Khidir & Zailani. (2011). Drivers on the reverse logistics: evidence from Malaysian certified companies. *International Journal of Logistics Systems and Management, 10*(4), 431–443.

Kumar, D., Rahman, Z., & Chan, F. T. S. (2016). A fuzzy AHP and fuzzy multi-objective linear programming model for order allocation in a sustainable supply chain : A case study, *3052*(February). https://doi.org/10.1080/0951192X.2016.1145813

Lai, C., Chiu, C., Yang, C. et al. (2010). The Effects of Corporate Social Responsibility on Brand Performance: The Mediating Effect of Industrial Brand Equity and Corporate Reputation. *J Bus Ethics 95*, 457–469. https://doi.org/10.1007/s10551-010-0433-1.

Leclerc, Y., & Partners, W. M. (2013). Sustainability and the supply chain : How to reduce cost and save the environment, https://www.manufacturing.net/operations/article/13057048/sustainability-and-the-supply-chain-how-to-reduce-cost-and-save-the-environment [Accessed on November 11, 2019].

Loosemore, M., & Lim, B. T. H. (2017). Linking corporate social responsibility and organizational performance in the construction industry. *Construction Management and Economics, 35*(3), 90–105. https://doi.org/10.1080/01446193.2016.1242762

Macchion, L., Moretto, A., Caniato, F., Caridi, M., Danese, P., Spina, G., et al. (2017). Improving innovation performance through environmental practices in the fashion industry: The moderating effect of internationalisation and the influence of collaboration. *Production Planning & Control, 28*(3), 190–201. https://doi.org/10.1080/09537287.2016.1233361

Meixell, M. J., & Luoma, P. (2015). International Journal of Physical Distribution & Logistics Management Article information. *International Journal of Physical Distribution & Logistics Management, 45*(1/2), 69–89. https://doi.org/10.1108/IJPDLM-05-2013-0155

Miroshnychenko, I., Barontini, R., & Testa, F. (2017). Green practices and financial performance: A global outlook. *Journal of Cleaner Production, 147*, 340–351. https://doi.org/10.1016/j.jclepro.2017.01.058

Mraihi, R., & Abid, M. (2013). Road transport-related energy consumption : Analysis of driving factors in Tunisia. *Energy Policy, 62*, 247–253. https://doi.org/10.1016/j.enpol.2013.07.007

Nitzl, C. (2016). The use of partial least squares structural equation modelling (PLS-SEM) in management accounting research: Directions for future theory development. *Journal of Accounting Literature, 37*, 19–35. https://doi.org/10.1016/j.acclit.2016.09.003

Nitzl, C., & Roldán, J. (2016). Mediation analysis in partial least squares path modeling: helping researchers discuss more sophisticated models. *Industrial Management & Data Systems*, 1–31. https://doi.org/10.1108/IMDS-07-2015-0302

Oh (2000). Diners' perceptions of quality, value, and satisfaction: A practical viewpoint. *The Cornell Hotel and Restaurant Administration Quarterly, 41*(3), 58–66.

Qin, J., Liu, X., & Pedrycz, W. (2017). An extended TODIM multi-criteria group decision making method for green supplier selection in an interval type-2 fuzzy environment. *European Journal of Operational Research, 258*(2), 626–638. https://doi.org/10.1016/j.ejor.2016.09.059

Rezaei, J., Nispeling, T., Sarkis, J., & Tavasszy, L. (2016). A supplier selection life cycle approach integrating traditional and environmental criteria using the best worst method. *Journal of Cleaner Production, 135*, 577–588. https://doi.org/10.1016/j.jclepro.2016.06.125

Rezaei, S., & Kheirkhah, A. (2017). A comprehensive approach in designing a sustainable closed-loop supply chain network using cross-docking operations. *Computational and Mathematical Organization Theory, 24*, 1–48. https://doi.org/10.1007/s10588-017-9247-3

Richter, N. F., Cepeda, G., Roldán, J. L., & Ringle, C. M. (2016). European management research using partial least squares structural equation modeling (PLS-SEM). *European Management Journal, 34*(6), 589–597. https://doi.org/10.1016/j.emj.2016.08.001

Rueda, X., Garret, R. D., Lambin, E. F., Garrett, R. D., & Lambin, E. F. (2017). Corporate investments in supply chain sustainability: Selecting instruments in the Agri-food industry. *Journal of Cleaner Production, 142*, 2480. https://doi.org/10.1016/j.jclepro.2016.11.026

Seles, B. M. R. P., de Sousa Jabbour, A. B. L., Jabbour, C. J. C., & Dangelico, R. M. (2016). The green bullwhip effect, the diffusion of green supply chain practices, and institutional pressures: Evidence from the automotive sector. *International Journal of Production Economics, 182*, 342–355. https://doi.org/10.1016/j.ijpe.2016.08.033

Shaharudin, M. R., Govindan, K., Zailani, S., Tan, K. C., & Iranmanesh, M. (2017). Product return management: Linking product returns, closed-loop supply chain activities, and the effectiveness of the reverse supply chains. *Journal of Cleaner Production, 149*, 1144–1156. https://doi.org/10.1016/j.jclepro.2017.02.133

Simão, L. E., Gonçalves, M. B., & Taboada Rodriguez, C. M. (2016). An approach to assess logistics and ecological supply chain performance using postponement strategies. *Ecological Indicators, 63*, 398–408. https://doi.org/10.1016/j.ecolind.2015.10.048

Song, W., Ming, X., & Liu, H.-C. (2017). Identifying critical risk factors of sustainable supply chain management: A rough strength-relation analysis method. *Journal of Cleaner Production, 143*, 100–115. https://doi.org/10.1016/j.jclepro.2016.12.145

Thöni, A., & Tjoa, A. M. (2015). Information technology for sustainable supply chain management : A literature survey management : A literature survey, *7575*(October). https://doi.org/10.1080/17517575.2015.1091950

Tseng, M. L., Lin, Y. H., Tan, K., Chen, R. H., & Chen, Y. H. (2014). Using TODIM to evaluate green supply chain practices under uncertainty. *Applied Mathematical Modelling, 38*(11–12), 2983–2995. https://doi.org/10.1016/j.apm.2013.11.018

Vinyes, E., Asin, L., Alegre, S., MuOz, P., Boschmonart, J., & Gasol, C. M. (2017). Life cycle assessment of apple and peach production, distribution, and consumption in Mediterranean fruit sector. *Journal of Cleaner Production, 149*, 313–320. https://doi.org/10.1016/j.jclepro.2017.02.102

Walker, H., Di Sisto, L., & McBain, D. (2008). Drivers and barriers to environmental supply chain management practices: Lessons from the public and private sectors. *Journal of Purchasing and Supply Management, 14*(1), 69–85.

Weerawardena, J., Mort, G. S., Salunke, S., Knight, G., & Liesch, P. W. (2014). The role of the market sub-system and the socio-technical sub-system in innovation and firm performance: A dynamic capabilities approach. *Journal of the Academy of Marketing Science., 43*, 221. https://doi.org/10.1007/s11747-014-0382-9

Xia, X. H., Chen, B., Wu, X. D., Hu, Y., Liu, D. H., & Hu, C. Y. (2017). Coal use for the world economy: Provision and transfer network by multi-region input-output analysis. *Journal of Cleaner Production, 143*, 125–144. https://doi.org/10.1016/j.jclepro.2016.12.142

Yang, D., & Xiao, T. (2017). Pricing and green level decisions of a green supply chain with governmental interventions under fuzzy uncertainties. *Journal of Cleaner Production, 149*, 1174–1187. https://doi.org/10.1016/j.jclepro.2017.02.138

Yenipazarli, A. (2016). To collaborate or not to collaborate : Prompting upstream eco-efficient innovation in a supply chain. *European Journal of Operational Research, 260*, 571–587. https://doi.org/10.1016/j.ejor.2016.12.035

Zhang, Y. H., & Wang, Y. (2017). The impact of government incentive on the two competing supply chains under the perspective of corporation social responsibility : A case study of the photovoltaic industry. *Journal of Cleaner Production, 154*, 102–113. https://doi.org/10.1016/j.jclepro.2017.03.127

Zhao, R., Liu, Y., Zhang, N., & Huang, T. (2017). An optimization model for green supply chain management by using a big data analytic approach. *Journal of Cleaner Production, 142*, 1085–1097. https://doi.org/10.1016/j.jclepro.2016.03.006

Zhu, Q., & Sarkis, J. (2004). Relationships between operational practices and performance among early adopters of green supply chain management practices in Chinese manufacturing enterprises. *Journal of Operations Management, 22*, 265–289.

Chapter 4
Statistical Analyses of Green Supply Chain Management

4.1 Survey Response Analysis

The responses of the survey were collected during the summer and winter holidays of 2016. The survey has been conducted in Pakistan to analyze the barriers and motivating components of the green supply chain in the manufacturing firms of Pakistan. The primary respondents of our survey were supply chain managers, directors, and supply chain professionals to maintain the data reliability and credibility of the surveys. First of all, we provide a brief introduction and purpose of this questionnaire to the respondents for realistic and trust gaining reasons. Then, the filled questionnaires are put into Excel files for coding and tagging analyzed accordingly to examine the validity and reliability of the data, as well as check the stipulated hypothesis.

4.1.1 Survey Response Analysis

The respondents came from the manufacturing firms, and their businesses are related to manufacturing different products. The respondents are mainly divided into three different positions, such as supply chain director, supply chain manager, and supply chain professional/officer. The detail of the respondents' demographic has been shown in Table 4.1.

Further, we have broken the demographic profile into graphical representation for a clear picture and understanding. Besides, we will discuss one by one each graph in detail. In this survey total, 415 forms are successfully included in the analysis for testing our research hypothesis. Figure 4.1 shows the respondents' job title, and here we can see that 21% of questionnaire forms are filled by supply chain directors, 47 forms filled by supply chain professionals, and 32% forms filled by supply chain managers.

S. A. R. Khan, *The Critical Success Factors of Green Supply Chain Management in Emerging Economies*, EAI/Springer Innovations in Communication and Computing,
https://doi.org/10.1007/978-3-030-42742-9_4

Table 4.1 Demographic profile

Demographics	Number of respondents	%
Job title		
Supply chain professionals	197	47
Supply chain managers	134	32
Supply chain directors	84	21
Industry category		
Chemical manufacturing	139	33
Food manufacturing	18	5
Fabricated metal product manufacturing	48	11
Electronic product manufacturing	12	3
Machinery manufacturing	13	3
Tobacco product manufacturing	6	2
Furniture manufacturing	9	2
Coal and petroleum product manufacturing	8	2
Paper manufacturing	34	8
Textile mills	86	21
Beverage product manufacturing	6	1
Transportation equipment manufacturing	13	3
Wood product manufacturing	11	3
Plastic and rubber manufacturing	9	2
Apparel manufacturing	3	1
Age of firm		
Less than 10 years	37	9
10–20 years	167	40
21–30 years	109	26
More than 30 years	102	25
Sales (millions US$)		
Less than $10	76	18
$11–50	90	22
$51–100	118	28
$101–200	75	18
More than $200	56	14
Number of employees		
Less than 50 workers	63	15
50–100 workers	175	42
101–200 workers	138	33
201–500 workers	23	6
More than 500 workers	16	4

In Fig. 4.2, we can see the industry category, which has been included in these research surveys. The number of questionnaire filled by chemical manufacturers was 139, textile mills was 86, and paper manufacturers was 34, which is 33%, 21%, and 8% of the total questionnaire filled by firms, respectively.

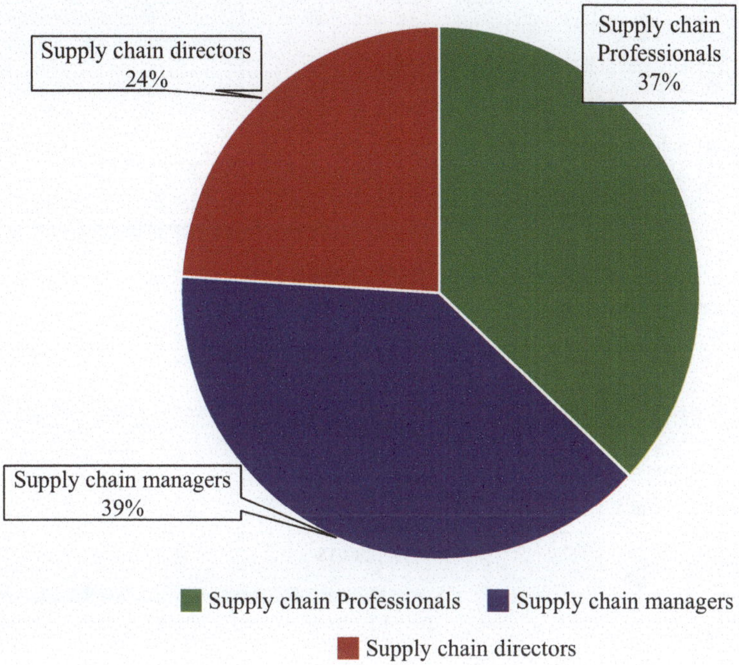

Fig. 4.1 Respondent's job title

Figure 4.3 illustrates the respondents firms' age, and we can quickly realize 167 firms are established from 10 to 20 years and 109 and 102 firms are 21 to 30 years and more than 30 years, respectively. In simple words, in our surveys, most firms have long experience in the market and have a credible position in local and international markets.

During surveys, firms' sales also collected from the questionnaire. Figure 4.4 displays the firms' sales in a million US$. The figure shows that 75 firms' sales are between the range of $101 and $200 million and 56 firms' sales are $200 million and above. In simple words, this analysis shows that most firms in this study are financially healthy.

Figure 4.5 shows the classification of the number of employees in the firms. One hundred thirty-eight firms have employees between the range of 101 and 200, 175 firms have the workforce between the range of 50 and 100, and 23 firms have their workforce between the range of 201 and 500. Most firms are large-scale manufacturing firms and/or medium-scale manufacturing firms, and minimal numbers of firms are from small manufacturing.

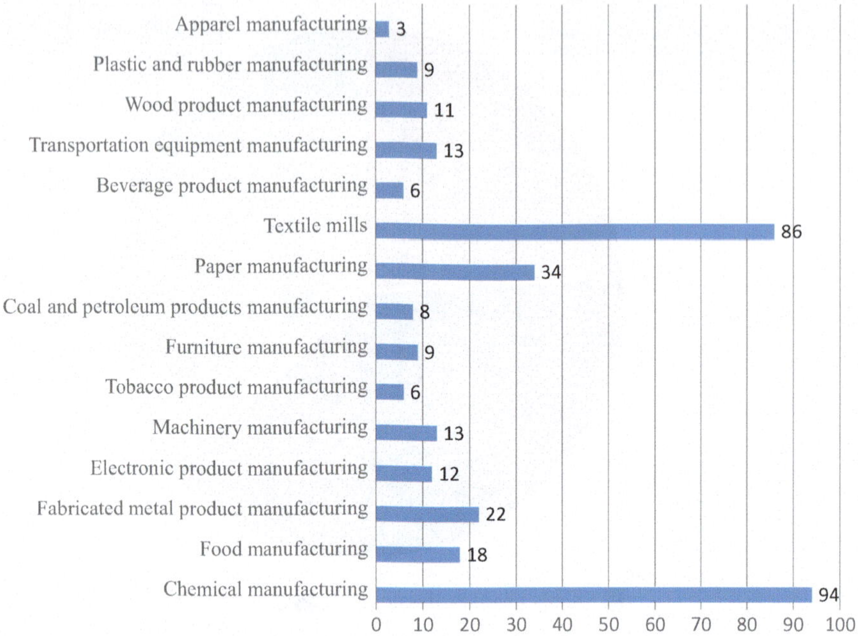

Fig. 4.2 Industry category

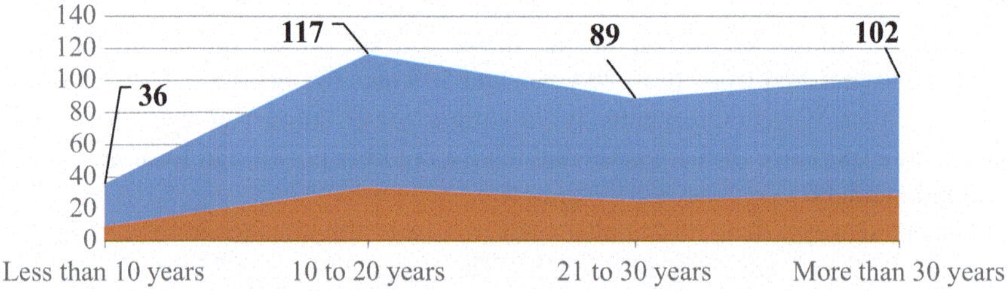

Fig. 4.3 Respondents firms' age

4.1.2 *Non-response and Response Bias*

As nonrespondents in the survey study could share some common characteristics and different from respondents, failure to attain their responses could directly or indirectly affect the research study findings. Due to these reasons, non-response bias was checked by *t*-tests in SPSS. As we have mentioned that some late respondents filled the forms after a couple of reminders, they could be well representative of nonrespondents. This is suggested by Armstrong and Overton (1977), an extrapolation method. Thus, answers of some late respondents who filled the survey after remainders and remaining respondents who filled the forms in first attempt without any remainders in respect of "supply chain director" ($t = 1.390$, $p < 0.15$), "supply

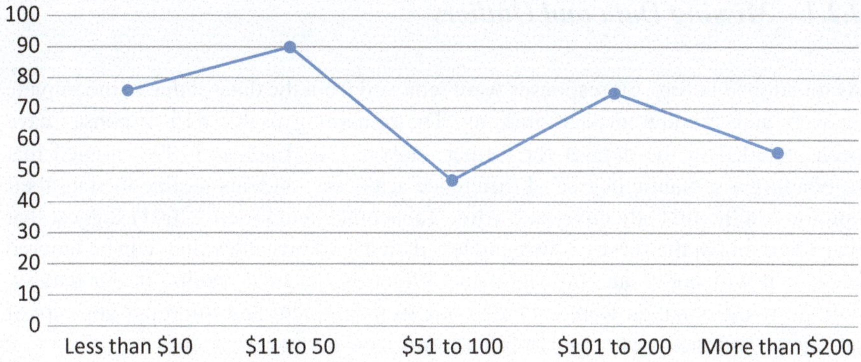

Fig. 4.4 Firms' sales in millions US$

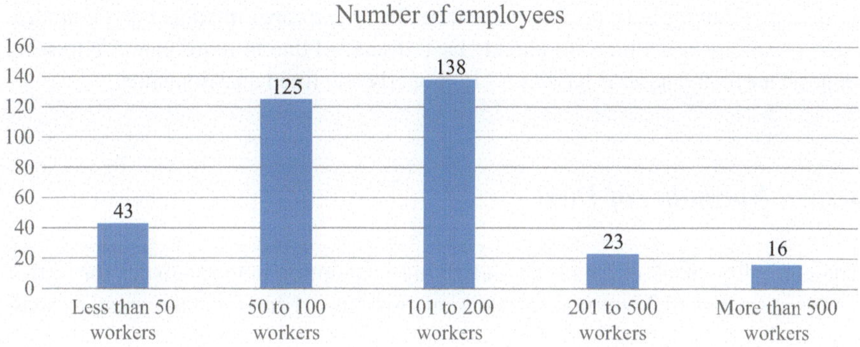

Fig. 4.5 Firms' respondent employees

chain manager" ($t = 0.624$, $p < 0.56$), and "supply chain supervisors" ($t = 0.758$, $p < 0.49$). The findings of the *t*-test show that there are no significant differences among early and late responses. Furthermore, because the late respondents took a couple of weeks due to busy schedules of meetings commitments or scheduled meetings and foreign visits to read the forms thoroughly and in-detailed understand the questionnaire and their contents. In simple words, the overall findings of *t*-tests revealed no significant differences regarding non-response bias in this research study.

4.2 Data Screening and Initial Analysis

In order to testify the suitability of using PLS path modeling in this research study, the degree to which the data attain fulfill psychometric assumptions was the first criteria. This included treatment of normality of data distribution missing data and common method bias, which all can have direct and/or indirectly influence on retaining the relevant techniques of data analysis.

4.2.1 Missing Data and Outliers

As mentioned before, 30 responses were removed from the dataset due to incomplete answers and/or some answers missing. The remaining usable 415 responses have been included in the dataset for further analysis. Al-Hindawe (1996) argued that responding a semantic degree of difference scale; respondents incline to state their opinions on the first adjective-pair, while Tabachnick and Fidell, (2001) suggest that if any variable in the survey is missing less than 5% so missing values can be ignored because it will not create any significant effect on the final results. In our dataset, only 30 questionnaires were excluded due to incomplete and improper answers of questions, and the remaining 415 filled forms have no missing values.

According to Hair, Anderson, Tatham, and Black (1998), the large sample size of 80 and higher, a standardized value of score ±3, can be suitable to be used as a standard or benchmark to recognize the outliers. In our dataset, some 18 responses were found as an outlier, but these 18 responses are from the 30 improper filled and/or incomplete filled questionnaire, which we already have discarded due to more than 5% missing values. Our final dataset is based on 415 properly and filled questionnaires.

4.2.2 Normality of Data

The normality check of data is an underlying assumption in the statistics for further analysis, such as SEM (structural equation modeling), Hierarchical regression analysis, and logistic regressions. For this reason, when assumptions of normality are violated, then alternative (Henseler, Ringle, & Sinkovics, 2009) statistical techniques/methods should be used. According to the Kline (2005) explain that if the kurtosis scores outside of +2 to −2 times its std. Error and skewness rating outside +1 to −1 time its std. An error has the potential to restrict the analysis of data and interpretation of subsequent results. In this study, all the normality of data assumptions has been fulfilled, and this deviation from assumption of normality is a compelling reason for adopting the PLS-SEM path (Henseler et al., 2009) modeling in this research. Furthermore, the original sample size for PLS-SEM is 200 and above (Garver & Mentzer, 1999; Hoelter, 1983). In our research study, the sample size is 415, so using PLS-SEM is not a barrier for us. Concerning the size of the sample and its relevant power, the suggestion of Cohen (1992) has been adopted. Here, for a sample size of 415, loading of a factor of 0.55 or higher is countably significant.

4.2.3 Common Method Bias

The common bias method has been considered a key reason for the error in measurement and also a severe threat to the validity of the model (Podsakoff & Organ, 1986). This research has used Harman's one-factor test to measure the effect of this bias (Podsakoff, Mackenzie, Lee, & Podsakoff, 2013). An EFA (exploratory factor analysis) was adopted using principal axis factor with extracted based on greater than one eigenvalues. The solutions of un-rotated did not create a factor, proposing that common-method variance did not occur to be a significant threat.

4.2.4 PLS-SEM

In the literature review, a limited number of researchers used PLS-SEM because it is a new method to resolve complex primary data analysis. This article will be very beneficial for the researchers to understand how to use PLS-SEM and interpret the results. PLS-SEM has experienced increasing dissemination in a variety of fields in recent years with normal and non-normal data, small and large sample sizes, and the use of formative indicators being the most prominent reasons for its application. Recent methodological research has extended PLS-SEM methodological toolbox to accommodate more complex model structures or handle data inadequacies such as heterogeneity.

The researcher adopts PLS-SEM for the quantitative part of this study. The PLS-SEM is a collection of statistical techniques that allow a set of relationships between one or more exogenous variables, be it continuous or discrete, to be examined. PLS-SEM is a modeling approach to maximizing the explained variance of the dependent constructs (Hair, Christian, & Marko, 2011). It is also efficient for theory building, theory testing, and exploratory analysis (Lowry and Gaskin, 2014). Furthermore, the following are the key benefits of PLS-SEM:

- Useful for preliminary theory building and causal-predictive analysis.
- PLS-SEM is a suitable method and works better in both cases (small sample size and big sample size).
- PLS-SEM is a causal modeling method that also enhances the validity and reliability of the results.
- It does not have the capability for comparison of the observed and proposed covariance matrices.
- Reliable for exploratory analysis, which is required for theory building.
- It can be used for both confirmatory and exploratory analyses.
- Avoids indeterminacy by composing constructs from factor scores and using these in subsequent calculations, yielding explicit factor scores.
- It can be used in models with formative (or a mix of reflective and formative) indicators.
- It is robust to violations of multivariate normal distributions.

The main advantage that I have found is that PLS-SEM makes fewer assumptions about how the data are distributed and is more robust to smaller sample size limitations and offers in-depth analysis. PLS path modeling also flexibly allows you to specify both formative and reflective constructs (latent variables) in one model.

However, the two (CB-SEM and PLS-SEM) are different and are useful for different research purposes and questions. Think of PLS path modeling as good for prediction (which factor predicts what?) and/or hypothesis generation (Hair, 2014).

4.3 Measurement Refinement and Initial Analysis

For validation and reliability purpose, a two-step model has been built (Anderson & Gerbing, 1988; Hair Jr, Black, Babin, & Anderson, 2006; Schumacker & Lomax, 2004). In the first step, we did validate and purify the measures through EFA for unchecked new scales. The software of the PLS path modeling application has been used to test and build the structural model.

4.3.1 Reliability and Validity of Measures

To validate the construct of latent proposed in the model, it has been checked through the EFA. Table 4.2 has showed the reliability of all the questions including IEM (internal environmental management); GIS (green information systems); CA (competitive edge); ENP (environmental performance); ECP (economic performance); ED (ecological design); GD (green distribution); GM (green manufacturing); GT (green transportation); and ORP (organizational performance).

In the research of natural sciences, the recommended reliability (value of Cronbach's alpha) should be a minimum of 70%. On the other hand, in the study of social science research, 60% reliability is also considered satisfactory and well acceptable in some circumstances for further analysis (Nunnally, 1978), while in our research study, the minimum cutoff value of reliability is 70% and above for maintaining the higher reliability and extract reliable results. As per the results of reliability and validity, we can continue our further analysis to test our research hypothesis.

Table 4.3 shows the construct reliability in our research study. The reliability of CA (competitive edge) constructs is 78%, ECP (economic performance) 74%, ED (ecological design) 81%, ENP (environmental performance) 78%, GD (green distribution) 78%, GIS (green information systems) 77%, GM (green manufacturing) 83%, GT (green transportation) 76%, IEM (internal environmental management) 72%, and ORP (organizational performance) 86%. In simple words, the reliability of all constructs is more than 70%; we can continue our further analysis to measure and test our main hypothesis.

Figure 4.6 shows the reliability of constructs, and the acceptable value of reliability in the social sciences is 40–50% (Nunnally, 1978), while in our research study, minimum reliability value is 57%. It means that we can continue the further analysis to test our study hypothesis.

Table 4.2 Reliability (Cronbach's alpha)

Questionnaire reliability	Cronbach's alpha
Green information systems	
GIS1	0.775
GIS2	0.771
GIS3	0.773
Green distribution	
GD1	0.743
GD2	0.814
GD3	0.740
GD4	0.710
Green transportation	
GT1	0.721
GT2	0.771
GT3	0.731
GT4	0.775
Ecological design	
ED1	0.791
ED2	0.793
ED3	0.831
Green manufacturing	
GM1	0.864
GM2	0.871
GM3	0.773
Economic performance	
ECP1	0.764
ECP2	0.732
Competitive edge	
CA1	0.743
CA2	0.773
CA3	0.793
Organizational performance	
ORP1	0.895
ORP2	0.871
ORP3	0.823
Internal environmental management	
IEM1	0.762
IEM2	0.708
IEM3	0.704
IEM4	0.732
Environmental performance	
ENP1	0.731
ENP2	0.768
ENP3	0.794

Table 4.3 Reliability and validity

	Cronbach's Alpha	Composite reliability	Average variance extracted (AVE)
CA	0.781	0.872	0.694
ECP	0.745	0.860	0.754
ED	0.817	0.844	0.646
ENP	0.778	0.821	0.609
GD	0.777	0.820	0.534
GIS	0.772	0.855	0.665
GM	0.831	0.846	0.655
GT	0.764	0.779	0.543
IEM	0.723	0.860	0.608
ORP	0.862	0.872	0.694

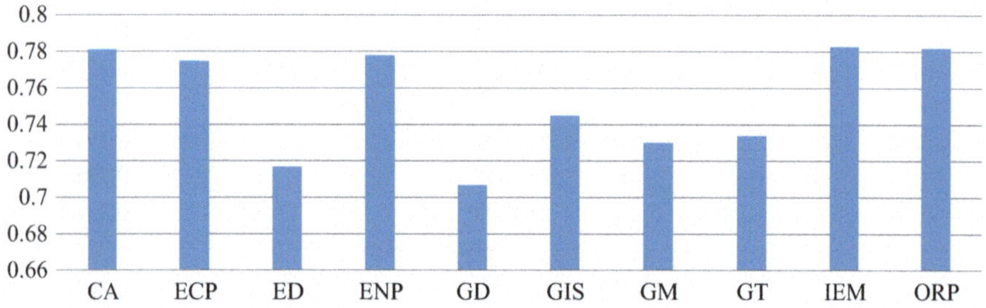

Fig. 4.6 Reliability of constructs

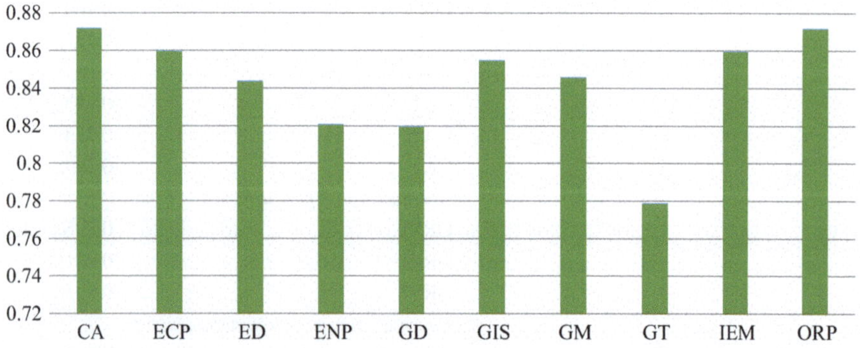

Fig. 4.7 Composite reliability

Figures 4.7 and 4.8 show the composite reliability and average variance extracted, respectively. The convergent validity of the measurement model can be assessed by the average variance extracted (AVE) and composite reliability (CA), and the minimum criteria for composite value is 70%, and the average variance extracted value is 50%, and in our study both requirements have been fulfilled successfully.

Table 4.4 shows the descriptive statistics construct wise of endogenous and exogenous variables. The table indicates that all of the constructs have significant positive values, which confirm the positive and robust green supply chain into the

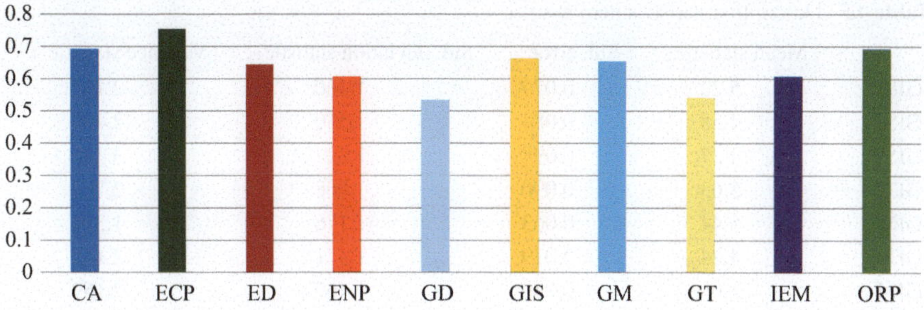

Fig. 4.8 Average variance extracted

Table 4.4 Descriptive statistics construct wise

	Mean statistic	Std. error	Std. deviation	Variance
GIS	14.0785	0.21315	3.95340	15.629
GD	20.3866	0.22703	4.21077	17.731
GT	16.8169	0.22594	4.19060	17.561
ED	13.9622	0.18665	3.46179	11.984
GM	13.3430	0.20477	3.79793	14.424
ECP	11.0291	0.11199	2.07702	4.314
CA	13.8663	0.20043	3.71736	13.819
ORP	14.3343	0.20305	3.76595	14.182
IEM	17.9797	0.21003	3.89545	15.175
ENP	9.8430	0.16039	2.97488	8.850

Pakistani manufacturing firms with healthier economic policies. Further, we have distributed construct-wise descriptive statistics into item-wise descriptive statistics for a more visible picture (Table 4.5).

Table 4.6 shows the correlational matrix and identify that green practices, namely, GIS (green information systems), ENP (environmental performance), IEM (internal environmental management), ECP (economic performance), CA (competitive edge), GM (green manufacturing), ED (Ecological design), GT (green transportation), and GD (green distribution), have positive correlation with organizational performance. The green practices are a significant contributor to improve firms' performance and reduce carbon emissions and energy consumption, while green practices are growing in trend, and many developed and developing countries are adopting green practices to improve their firm's performance. On the other hand, the government also encourages firms to adopt green practices in terms of tax exemptions, low tariff, and loan offer on the lowest interest rate.

Green practices have a positive and significant relationship with organizational performance. It means organizational performance significantly improved by the adoption of green practices in the firms. On the other hand, green practices also reduce carbon emission and hazardous and toxic chemicals used in manufacturing processes, while greater environmental sustainability and healthy economic activities can be achieved through green practices and foreign direct investment is also attracted by country economic sustainability.

Table 4.5 Descriptive statistics item wise

	Mean statistic	Std. error	Std. deviation statistic	Variance statistic
GIS1	5.22	0.079	1.460	2.132
GIS2	4.58	0.084	1.550	2.401
GIS3	4.27	0.098	1.827	3.336
GD1	5.63	0.090	1.668	2.782
GD2	5.44	0.063	1.176	1.384
GD3	4.24	0.079	1.471	2.163
GD4	5.08	0.079	1.469	2.159
GT1	3.08	0.089	1.654	2.735
GT2	3.99	0.080	1.481	2.192
GT3	5.24	0.092	1.707	2.912
GT4	4.51	0.096	1.783	3.178
ED1	3.98	0.085	1.585	2.513
ED2	4.99	0.070	1.291	1.668
ED3	4.99	0.079	1.460	2.131
GM1	2.94	0.113	2.101	4.414
GM2	4.81	0.078	1.454	2.113
GM3	5.59	0.066	1.216	1.479
ECP1	5.98	0.053	.981	.962
ECP2	5.05	0.075	1.399	1.956
CA1	4.37	0.081	1.497	2.240
CA2	5.08	0.079	1.473	2.168
CA3	4.41	0.080	1.490	2.220
ORP1	5.01	0.077	1.424	2.029
ORP2	4.92	0.083	1.537	2.362
ORP3	4.40	0.084	1.556	2.421
IEM1	4.84	0.076	1.417	2.008
IEM2	6.03	0.055	1.020	1.040
IEM3	3.76	0.069	1.287	1.656
IEM4	3.35	0.071	1.314	1.727
ENP1	3.13	0.079	1.464	2.145
ENP2	3.88	0.066	1.227	1.505
ENP3	2.84	0.062	1.149	1.320

For the purpose that the items were calculated to index ten constructs: green information systems (GIS), internal environmental management (IEM), green distribution (GD), green transportation (GT), green manufacturing (GM), competitive edge (CA), economic performance (ECP), environmental performance (ENP), Ecological design (ED) (predictors), while organizational performance (ORP) (dependent variable).

The value of KMO (Kaiser-Meyer-Olkin) extent of sampling adequacy (0.914) presents that the sample is suitable, and overall questionnaire reliability is sufficient to run factor analysis. Furthermore, a significant result of Kaiser-Meyer-Olkin ($p < 0.05$) shows that the matrix is not an identity matrix. In short, these ten

Table 4.6 Correlational matrix

	GIS	GD	GTF	ED	GM	ECP	CA	ORP	IEM	ENP
GIS	1									
GD	0.653**	1								
GT	0.552**	0.596**	1							
ED	0.638**	0.609**	0.626**	1						
GM	0.580**	0.517**	0.449**	0.631**	1					
ECP	0.642**	0.465**	0.357**	0.600**	0.527**	1				
CA	0.621**	0.620**	0.537**	0.624**	0.517**	0.612**	1			
ORP	0.715**	0.628**	0.542**	0.702**	0.567**	0.651**	0.715**	1		
IEM	0.721**	0.597**	0.559**	0.697**	0.576**	0.728**	0.702**	0.886**	1	
ENP	0.685**	0.595**	0.595**	0.705**	0.587**	0.587**	0.692**	0.848**	0.783**	1

Table 4.7 KMO and Bartlett's test

Kaiser-Meyer-Olkin measure of sampling adequacy		0.914
Bartlett's test of sphericity	Approx. chi-square	9850.743
	df	496
	Sig.	0.000

constructs relate from each other sufficient to estimate a meaningful factor analysis. Table 4.7 shows the KMO (Kaiser-Meyer-Olkin) and Bartlett's test findings.

Table 4.8 shows the commonalities which indicate how much of the variance (i.e., the value of commonality which should be more than 50% will be considered for next stage of analysis) in the variables has been estimated for by the factors of extracted. For example, 91% of the variance in ORP1, 95% of the variance in ORP2, and 96% of the variance in ORP3 are accounted for, while 67%, 89%, and 90% of the variance are accounted for in ENP1, ENP2, and ENP3, respectively. Further, 71% of the variance in IEM1, 75% of the variance in IEM2, 97% of the variance in IEM3, and 97% of the variance in IEM4 are accounted for (see Table 4.8).

In Fig. 4.9, the component plot in rotated space shows there are ten factors in the model. The included factors in the model are IEM (internal environmental management), GIS (green information systems), ED (Ecological design), GD (green distribution), GT (green transportation), GM (green manufacturing), CA (competitive edge), ECP (economic performance), ENP (environmental performance), and ORP organizational performance, while these ten factors explained 77%, as presented in Table 4.9.

The value of eigen shows the number of extracted factors whose total should be equal to many items when they are exposed to the analysis of factors. Further, it shows all the extractable factors from the analysis along with eigenvalues. Mainly, the table of eigenvalue has been distributed into three sub-parts, i.e., extracted sums of squared loadings, initial eigenvalues, and rotation of sums of squared loadings (Table 4.10). For interpretation and analysis purposes, we are concerned with the portion of extracted sums of squared loadings. Here we can see that first factors contribute to 42.669% of the variance, while after the tenth component, the remaining is not significant to include in the further analysis (see Table 4.9).

Table 4.8 Commonalities

	Initial	Extraction
GIS1	1.000	0.638
GIS2	1.000	0.649
GIS3	1.000	0.670
GD1	1.000	0.720
GD2	1.000	0.664
GD3	1.000	0.704
GD4	1.000	0.651
GT1	1.000	0.670
GT2	1.000	0.645
GT3	1.000	0.797
GT4	1.000	0.954
ED1	1.000	0.695
ED2	1.000	0.878
ED3	1.000	0.784
GM1	1.000	0.775
GM2	1.000	0.605
GM3	1.000	0.879
ECP1	1.000	0.684
ECP2	1.000	0.719
CA1	1.000	0.790
CA2	1.000	0.691
CA3	1.000	0.640
ORP1	1.000	0.906
ORP2	1.000	0.952
ORP3	1.000	0.961
IEM1	1.000	0.710
IEM2	1.000	0.749
IEM3	1.000	0.970
IEM4	1.000	0.965
ENP1	1.000	0.673
ENP2	1.000	0.890
ENP3	1.000	0.961

Extraction method: principal component analysis

4.3.2 *Confirmatory Factor Analysis*

In this research study, to be consistent with the choice analysis of multivariate, CFA (confirmatory factor analysis) for all reflective constructs has been completed through using the software SmartPLS (Ringle, Wende, & Will, 2005). The analysis of CFA has been performed through the extracted scale from the exploratory factor analysis. The reliability of all the reflective scales has been measured, followed by an assessment of their discriminant and convergent validities. The proposed model has been built in the software SmartPLS for testing hypotheses. In the model, latent

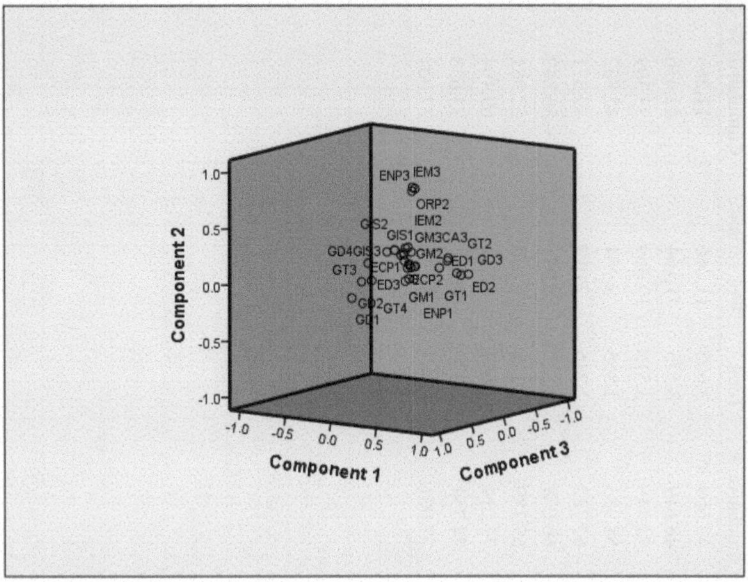

Fig. 4.9 Component plot in rotated space

constructs and respective measurement items are indicated through "circle shapes" and "rectangle shapes," respectively. These techniques of the hierarchical latent variable have been recommended by Wezels, Odekerken-Schroder, and van Oppen (2009).

As the model has been built properly in the software SmartPLS, essential statistics were calculated by the PLS algorithm. This complete process is usually called the outer measurement model. The results of the algorithm in loadings all are more than 70% (see Table 4.11). In our PLS-SEM model, we have included 32 items under the 10 constructs, and the details are as follows: IEM construct is measured through top management commitment towards GSCM, cross-functional cooperation for environmental improvement and environmental management systems, total quality environmental management and ISO 14001 certifications, and environmental compliance and auditing programs. GIS constructs include monitoring emission and waste production, improving decision-making of executives by highlighting sustainability problems, and reducing energy consumption and limiting carbon emissions. GD construct estimating distribution centers is working on environmental sustainability, and distribution centers are nearby customers, using shortest route for arriving on destination, and DC works on waste reduction and reducing electricity consumptions and uses eco-product packaging. GT construct covers reducing transportation inbound cost, using fuel-efficient vehicles to reduce emissions, vehicle maintenance, and reducing transportation outbound cost. ED construct includes design of products for reduced consumption of material/energy; design of products for reuse, recycle, and recovery of material or components parts; and design of products to avoid or reduce use of hazardous products and their manufacturing

Table 4.9 Total variance explained

Component	Initial eigenvalues			Extraction sums of squared loadings			Rotation sums of squared loadings		
	Total	% of variance	Cumulative %	Total	% of variance	Cumulative %	Total	% of variance	Cumulative %
1	13.654	42.669	42.669	13.654	42.669	42.669	3.605	11.266	11.266
2	1.968	6.150	48.819	1.968	6.150	48.819	3.573	11.164	22.430
3	1.802	5.630	54.449	1.802	5.630	54.449	3.510	10.969	33.399
4	1.451	4.534	58.983	1.451	4.534	58.983	2.865	8.954	42.352
5	1.197	3.739	62.722	1.197	3.739	62.722	2.246	7.019	49.371
6	1.061	3.316	66.038	1.061	3.316	66.038	2.237	6.991	56.362
7	1.008	3.151	69.189	1.008	3.151	69.189	2.118	6.620	62.981
8	1.007	3.026	72.214	1.007	3.026	72.214	2.084	6.513	69.494
9	1.004	2.539	74.754	1.004	2.539	74.754	1.339	4.184	73.678
10	1.001	2.247	77.001	1.001	2.247	77.001	1.063	3.323	77.001
11	0.809	2.216	79.218						
12	0.770	2.095	81.312						
13	0.594	1.856	83.168						
14	0.555	1.734	84.902						
15	0.516	1.613	86.515						
16	0.489	1.529	88.044						
17	0.446	1.394	89.438						
18	0.418	1.307	90.745						
19	0.405	1.266	92.011						
20	0.399	1.246	93.256						
21	0.374	1.168	94.424						
22	0.334	1.042	95.466						
23	0.317	0.989	96.456						
24	0.282	0.880	97.336						
25	0.264	0.825	98.161						

26	0.235	0.734	98.895						
27	0.147	0.459	99.354						
28	0.113	0.354	99.707						
29	0.042	0.131	99.838						
30	0.024	0.076	99.914						
31	0.017	0.055	99.968						
32	0.010	0.032	100.000						

Extraction method: principal component analysis

Table 4.10 Rotated component matrix[a]

	1	2	3	4	5	6	7	8	9	10
ORP1	0.873									
ORP2	0.858									
ORP3	0.765									
IEM1		0.792								
IEM2		0.864								
IEM3		0.803								
IEM4		0.75								
GT1			0.718							
GT2			0.85							
GT3			0.76							
GM1				0.704						
GM2				0.885						
GM3				0.917						
GIS1					0.724					
GIS2					0.87					
GIS3					0.845					
GD1						0.735				
GD2						0.81				
GD3						0.731				
GD4						0.737				
ENP1							0.705			
ENP2							0.84			
ENP3							0.87			
ED1								0.859		
ED2								0.873		
ED3								0.725		
ECP1									0.862	
ECP2									0.875	
CA1										0.78
CA2										0.866
CA3										0.851

Extraction method: principal component analysis
Rotation method: varimax with Kaiser normalization
[a]Rotation converged in 12 iterations

process. GM construct covers green practices in manufacturing processes and reducing fossil fuel energy consumption, decreasing toxic and hazardous chemicals in manufacturing processes, and using green materials in manufacturing. ECP construct calculates decrease in cost for energy consumption, decrease in cost of material purchasing, and decrease in fee of waste treatment and waste discharge. CA construct includes good quality and fewer defects, improved firm image and reputation, customers' positive feedback, repeat buying, fast delivery, and greater visibility. ENP construct covers reduction of air emissions, effluent waste and solid waste,

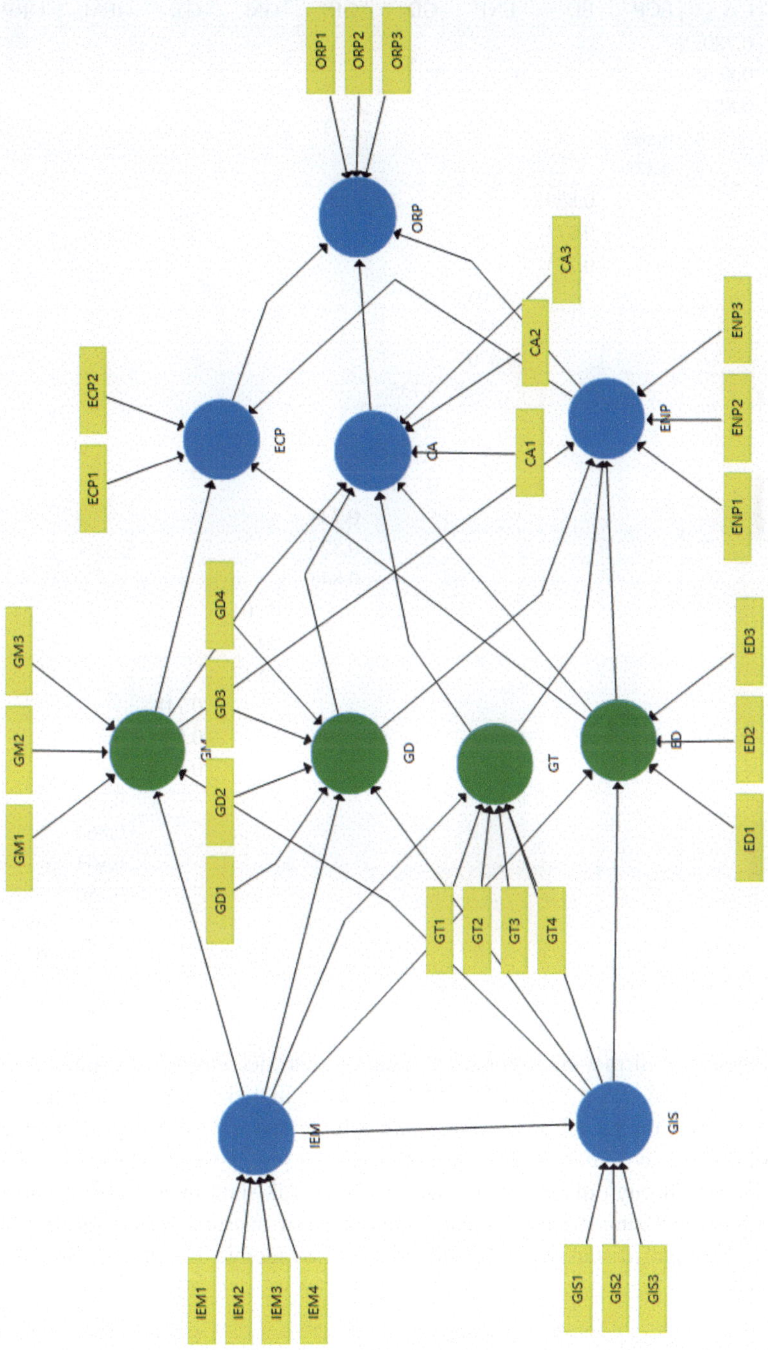

Fig. 4.10 Measurement model for testing

Table 4.11 Outer loading of indicators

	CA	ECP	ED	ENP	GD	GIS	GM	GT	IEM	ORP
CA1	0.780									
CA2	0.866									
CA3	0.851									
ECP1		0.862								
ECP2		0.875								
ED1			0.859							
ED2			0.873							
ED3			0.725							
ENP1				0.705						
ENP2				0.840						
ENP3				0.870						
GD1					0.735					
GD2					0.810					
GD3					0.731					
GD4					0.737					
GIS1						0.724				
GIS2						0.870				
GIS3						0.845				
GM1							0.704			
GM2							0.885			
GM3							0.917			
GT1								0.718		
GT2								0.850		
GT3								0.760		
IEM1									0.792	
IEM2									0.864	
IEM3									0.803	
IEM4									0.750	
ORP1										0.873
ORP2										0.858
ORP3										0.765

improvement in a firm's environmental situation, and decrease in consumption of hazardous/harmful, toxic materials. ORP construct includes avg. ROI (return on investment) over the past 3 years, avg. sales volume and in USD$ growth over the past 3 years, and avg. market share growth over the past 3 years. These data have been collected through questionnaire surveys from Pakistani manufacturing firms, while before finalizing the instrument, the content of instrument and validity has been checked and discussed with industry and academic researchers in detail (Fig. 4.10).

4.4 Model Validation

Before we go towards a testing model hypothesis, the validity of its measurement is also very important to be checked. However, for the construct of formative, the usually used analysis of validity is not appropriable. Therefore, Diamantopoulos and WinKlhofer's (2001) suggested checking first collinearity problems in the construct measurement. The formative measure with high collinearity shows that a construct of latent caused by multi-items has some dismissed items. The problem can be identified through VIF (variance inflation factor) values, while in our model, we have reflective all constructs measure. If VIF values are more significant than 5, so it means multicollinearity problems exist in the measurement model, and collinearity problems can distort the results of the measurement model. On the other hand, if weight values are negative, so it is also a symptom of the collinearity problem in the measurement model. Both tests, VIF and weight estimation, were conducted through using the software SmartPLS (see Tables 4.12 and 4.13 for weight values and VIF values, respectively).

In Table 4.12, outer weight values have been shown, and we can see that all values of outer weights are positive; it means that in our estimation model, there are no multicollinearity problems, and we can continue our further analysis.

In Table 4.13, variance inflation factor (VIF) values are smaller than 5, which shows that there are no multicollinearity problems in our estimation model.

4.4.1 Convergent Validity

The validity of convergent was assessed using CR (composite reliability) (Fornell & Larcker, 1981) and AVE (average variance extracted) scores. In this model estimation, CR is a more suitable measure of internal consistency as compared to Cronbach's alpha reliability, since it considers (Ma & Agarwal, 2007) the actual loadings while measuring indicators. In SEM-PLS of SmartPLS software, CR value higher than 70% is considered (Hair et al., 1998) acceptable for internal consistency. Furthermore, the value of average variance extracted 50% shows acceptable and robust convergent validity, as this means that more than 50% variation in a specific construct is enlightened (Chin & Newsted, 1999) by the stipulated indicators. Table 4.14 indicates internal consistency and convergent validity.

4.4.2 Discriminant Validity

The validity of discriminant is achieved if the square root of the average variance extracted of every construct is more significant than its bivariate association with other constructs and if a load of the indicator is higher (Chin, 1998; Compeau, Higgins & Huff, 1999) on their respective constructs when compared to remaining indicators (Table 4.15).

Table 4.12 Outer weight values

	CA	ECP	ED	ENP	GD	GIS	GM	GT	IEM	ORP
CA1	0.333									
CA2	0.440									
CA3	0.422									
ECP1		0.563								
ECP2		0.589								
ED1			0.441							
ED2			0.434							
ED3			0.365							
ENP1				0.291						
ENP2				0.484						
ENP3				0.480						
GD1					0.284					
GD2					0.374					
GD3					0.377					
GD4					0.340					
GIS1						0.362				
GIS2						0.455				
GIS3						0.405				
GM1							0.250			
GM2							0.483			
GM3							0.465			
GT1								0.380		
GT2								0.537		
GT3								0.428		
IEM1									0.272	
IEM2									0.344	
IEM3									0.346	
IEM4									0.317	
ORP1										0.434
ORP2										0.444
ORP3										0.313

The discriminant validity assessment has the objective to confirm that a construct of reflective has as much as strong associations with its constructs (Hair et al., 2017).

For the analysis of cross-loadings, we found that all indicators loaded highest in their respective factors (see Table 4.16), although there were no significant cross-loading cases found. In simple words, the model of our research study indicates the acceptable validity of discriminant. On the other hand, Henseler, Ringle, and Sarstedt (2015) argue that Fornell-Larcker criterion and cross-loading criteria are not very reliable to detect the discriminant validity problems, and they suggest to use HTMT (heterotrait-monotrait ratio of correlations) and confirmed that this test reliability is sufficient to process the model. If the value of the HTMT (heterotrait-monotrait ratio of correlations) value is below 0.90, discriminant validity has been established, and the model is reliable to process further (see Table 4.16).

Table 4.13 Inner VIF values

	CA	ECP	ED	ENP	GD	GIS	GM	GT	IEM	ORP
CA										2.249
ECP										1.783
ED	2.747	2.378		2.747						
ENP		2.122								2.314
GD	2.122			2.122						
GIS			2.132		2.132		2.132	2.132		
GM	1.886	1.976		1.886						
GT	2.459			2.459						
IEM			2.132		2.132	1.000	2.132	2.132		
ORP										

Table 4.14 Convergent validity of constructs

	Composite reliability	Average variance extracted (AVE)
CA	0.872	0.694
ECP	0.860	0.754
ED	0.844	0.646
ENP	0.821	0.609
GD	0.820	0.534
GIS	0.855	0.665
GM	0.846	0.655
GT	0.779	0.543
IEM	0.860	0.608
ORP	0.872	0.694

Table 4.15 Discriminant validity by Fornell-Larcker criterion

	CA	ECP	ED	ENP	GD	GIS	GM	GT	IEM	ORP
CA	0.833									
ECP	0.606	0.869								
ED	0.626	0.595	0.804							
ENP	0.716	0.621	0.698	0.781						
GD	0.640	0.479	0.636	0.621	0.731					
GIS	0.634	0.652	0.637	0.710	0.667	0.815				
GM	0.542	0.610	0.670	0.619	0.536	0.622	0.809			
GT	0.596	0.429	0.707	0.656	0.685	0.626	0.527	0.737		
IEM	0.724	0.743	0.699	0.837	0.610	0.729	0.638	0.629	0.780	
ORP	0.731	0.658	0.696	0.919	0.642	0.723	0.609	0.615	0.894	0.833

Table 4.16 Cross-loading

	CA	ECP	ED	ENP	GD	GIS	GM	GT	IEM	ORP
CA1	**0.780**	0.467	0.468	0.487	0.470	0.461	0.398	0.432	0.499	0.473
CA2	**0.866**	0.571	0.527	0.661	0.569	0.578	0.491	0.531	0.677	0.690
CA3	**0.851**	0.473	0.565	0.624	0.553	0.536	0.458	0.518	0.616	0.641
ECP1	0.467	**0.862**	0.499	0.512	0.430	0.576	0.536	0.387	0.656	0.559
ECP2	0.584	**0.875**	0.535	0.566	0.402	0.556	0.524	0.358	0.636	0.583
ED1	0.600	0.454	**0.859**	0.615	0.531	0.520	0.473	0.608	0.567	0.599
ED2	0.510	0.499	**0.873**	0.594	0.582	0.522	0.538	0.709	0.606	0.582
ED3	0.383	0.487	**0.663**	0.462	0.408	0.496	0.625	0.359	0.507	0.490
ENP1	0.419	0.274	0.515	**0.605**	0.398	0.408	0.413	0.549	0.408	0.398
ENP2	0.617	0.573	0.530	**0.840**	0.545	0.587	0.527	0.480	0.687	0.848
ENP3	0.616	0.549	0.608	**0.870**	0.503	0.641	0.508	0.550	0.804	0.819
GD1	0.391	0.289	0.303	0.327	**0.735**	0.439	0.321	0.431	0.357	0.379
GD2	0.535	0.357	0.449	0.480	**0.810**	0.509	0.389	0.504	0.435	0.487
GD3	0.484	0.379	0.685	0.527	**0.631**	0.487	0.464	0.534	0.513	0.515
GD4	0.432	0.354	0.365	0.442	**0.737**	0.496	0.366	0.509	0.451	0.464
GIS1	0.399	0.620	0.503	0.513	0.385	**0.724**	0.493	0.365	0.610	0.540
GIS2	0.631	0.545	0.567	0.656	0.656	**0.870**	0.557	0.572	0.626	0.655
GIS3	0.500	0.443	0.486	0.557	0.566	**0.845**	0.469	0.576	0.551	0.566
GM1	0.322	0.217	0.347	0.328	0.354	0.333	**0.584**	0.333	0.295	0.319
GM2	0.518	0.576	0.568	0.586	0.482	0.582	**0.885**	0.482	0.592	0.571
GM3	0.455	0.597	0.665	0.546	0.462	0.554	**0.917**	0.452	0.598	0.545
GT1	0.373	0.223	0.528	0.418	0.309	0.358	0.324	**0.688**	0.384	0.357
GT2	0.496	0.403	0.639	0.597	0.485	0.517	0.499	**0.850**	0.568	0.552
GT3	0.440	0.298	0.383	0.412	0.718	0.495	0.317	**0.660**	0.415	0.429
IEM1	0.422	0.545	0.449	0.444	0.397	0.476	0.446	0.418	**0.692**	0.444
IEM2	0.657	0.748	0.568	0.683	0.513	0.612	0.583	0.481	**0.864**	0.719
IEM3	0.621	0.550	0.597	0.859	0.505	0.631	0.497	0.539	**0.803**	0.826
IEM4	0.532	0.467	0.554	0.585	0.479	0.539	0.455	0.516	**0.750**	0.761
ORP1	0.635	0.580	0.551	0.829	0.574	0.603	0.536	0.493	0.696	**0.873**
ORP2	0.645	0.579	0.621	0.850	0.534	0.653	0.518	0.536	0.814	**0.858**
ORP3	0.538	0.475	0.578	0.578	0.496	0.544	0.467	0.521	0.736	**0.765**

Table 4.17 displays the results of HTMT (heterotrait-monotrait ratio), and all values in these findings should be less than 0.9 (Henseler, Ringle, & Sarstedt 2015). The results show that all the values are smaller than 0.9 values, and it confirmed that results are excellent and appropriate to continue further analysis.

Table 4.17 Heterotrait-monotrait ratio (HTMT)

	CA	ECP	ED	ENP	GD	GIS	GM	GT	IEM	ORP
CA										
ECP	0.830									
ED	0.829	0.864								
ENP	0.765	0.884	0.615							
GD	0.844	0.685	0.866	0.877						
GIS	0.816	0.729	0.879	0.798	0.899					
GM	0.704	0.820	0.692	0.866	0.736	0.823				
GT	0.881	0.674	0.808	0.607	0.658	0.794	0.797			
IEM	0.608	0.722	0.833	0.810	0.806	0.895	0.811	0.792		
ORP	0.817	0.899	0.835	0.520	0.850	0.694	0.784	0.791	0.813	

Table 4.18 *R*-square and adjusted *R*-square

	R-square	*R*-square adjusted
CA	0.510	0.504
ECP	0.481	0.477
ED	0.523	0.520
ENP	0.586	0.582
GD	0.478	0.475
GIS	0.531	0.530
GM	0.459	0.456
GT	0.455	0.452
ORP	0.863	0.862

4.4.3 *Quality Criteria of Model*

The *R*-square value is a statistical measure of how near the data are to be fitted in the line of regression. It is well-known as the coefficient of determination or the coefficient of multiple determination for regression analysis. The simple definition of *R*-square is relatively straightforward; it is the response percentage of the variable variation that is explained by a linear model.

Usually, the higher the *R*-square means the results are more reliable and useful, but in the primary research, the *R*-square value of 30% is called sufficient and reliable to interpret the results. Table 4.18 indicates the values of *R*-square and adjusted *R*-square of each construct. Fig. 4.11 indicates the *R*-square chart, and Fig. 4.12 shows the adjusted *R*-square graph for visual representation.

Table 4.19 indicates the *f*-square values; as per Cohen (1988) the effect size is calculated if an exogenous latent variable has a substantial impact on endogenous latent variables. It is estimated as the increase in *R*-square of the latent variable to which the path is connected, related to the latent variable (Chin, 1998) proportion of unexplained variance.

Table 4.19 The *F*-square values

	CA	ECP	ED	ENP	GD	GIS	GM	GT	ORP
CA									0.415
ECP									0.516
ED	0.037	0.257		0.172					
ENP		0.887							1.862
GD	0.301			0.133					
GIS			0.073		0.202		0.097	0.110	
GM	0.224	0.086		0.162					
GT	0.132			0.484					
IEM			0.246		0.630	1.132	0.134	0.117	
ORP									

The values of *f*-square between the range of 0.020 and 0.150 have a small effect; between the range of 0.150 and 0.350 have a medium effect; and between the range of 0.350 and above have a significant effect on an endogenous latent variable (Gefen, Straub, & Boudreau, 2000).

In Table 4.19, we can see that all the values of *f*-square are more significant than 0.020, and most values are around 0.350 and above, which shows the strong effect on the constructs and validity of the model.

4.5 Structural Model Evaluation

This research study model contains ten latent variables that need to be analyzed using a suitable statistical method that captures the estimation of the scores. Therefore, an SEM-PLS algorithm was performed on this model to measure the loadings of the variables and their weight (as mentioned in Tables 4.11 and 4.12). The SM (structural model) than estimated using the bootstrapping procedure (Chin, 1998). The findings are provided in the following paragraph.

4.5.1 Structural Model Evaluation

After applying the bootstrapping technique, the findings are given in Fig. 4.13. As per the *t*-values and sig value estimates from this bootstrapping process. The path CA > ORP, ECP > ORP, ED > ECP, ED > ENP, ENP > ECP, ENP > ORP, GD > CA, GD > ENP, GIS > ED, GIS > GD, GIS > GM, GIS > GT, GM > ECP, GM > ENP, GT > ENP, IEM > ED, IEM > GD, IEM > GIS, IEM > GM, IEM > GT, ED > CA is statistically and positively significant at the 1% level of confidence, while GM > CA is statistically and positively significant at the 5% level of confidence and GT > CA is statistically and positively significant at the 10% level of confidence. Figure 4.14 and Table 4.20 show the coefficient values of exogenous and endogenous variables.

Table 4.20 Results of hypotheses testing

Hypothesis	Effect	Coefficient	*T* statistics	*p* values	Remarks
1	IEM – > GIS	0.729	24.709	0.000	Accepted on 1% confidence level
2	IEM – > GM	0.393	6.11	0.000	Accepted on 1% confidence level
3	IEM – > GD	0.265	4.425	0.000	Accepted on 1% confidence level
4	IEM – > GT	0.368	6.036	0.000	Accepted on 1% confidence level
5	IEM – > ED	0.5	8.455	0.000	Accepted on 1% confidence level
6	GIS – > GM	0.335	5.293	0.000	Accepted on 1% confidence level
7	GIS – > GD	0.474	7.819	0.000	Accepted on 1% confidence level
8	GIS – > GT	0.357	6.133	0.000	Accepted on 1% confidence level
9	GIS – > ED	0.273	4.398	0.000	Accepted on 1% confidence level
10	GM – > ECP	0.297	5.365	0.000	Accepted on 1% confidence level
11	GM – > CA	0.144	2.3	0.022	Accepted on 5% confidence level
12	GM – > ENP	0.219	4.111	0.000	Accepted on 1% confidence level
13	GD – > CA	0.325	4.736	0.000	Accepted on 1% confidence level
14	GD – > ENP	0.169	2.831	0.005	Accepted on 1% confidence level
15	GT – > CA	0.139	1.751	0.081	Accepted on 10% confidence level
16	GT – > ENP	0.222	3.85	0.000	Accepted on 1% confidence level
17	ED – > ECP	0.178	2.693	0.007	Accepted on 1% confidence level
18	ED – > CA	0.225	3.12	0.002	Accepted on 1% confidence level
19	ED – > ENP	0.287	4.478	0.000	Accepted on 1% confidence level
20	ENP – > ECP	0.313	5.064	0.000	Accepted on 1% confidence level
21	ENP – > ORP	0.769	26.829	0.000	Accepted on 1% confidence level
22	CA – > ORP	0.113	3.557	0.000	Accepted on 1% confidence level
23	ECP – > ORP	0.112	4.058	0.000	Accepted on 1% confidence level

Furthermore, the variances shown in the circles are deemed. Figure 4.14 displays the coefficient values as a graphical representation to easily understand. Fifty-three percent of green information systems have been explained by the internal environmental management component. The 47.8% green distribution, 46% green manufacturing, 52% ecological design, and 46% green transportation have been explained by internal environmental management and green information systems exogenous variables. On the other hand, 48% economic performance has been explained by green manufacturing, environmental performance, and ecological design; 51% competitive edge has been explained by green distribution, green manufacturing, ecological design, and green transportation; 59% environmental performance has been explained by green distribution, ecological design, green manufacturing, and green transportation, and finally 87% organizational performance has been explained by competitive edge, economic performance, and environmental performance.

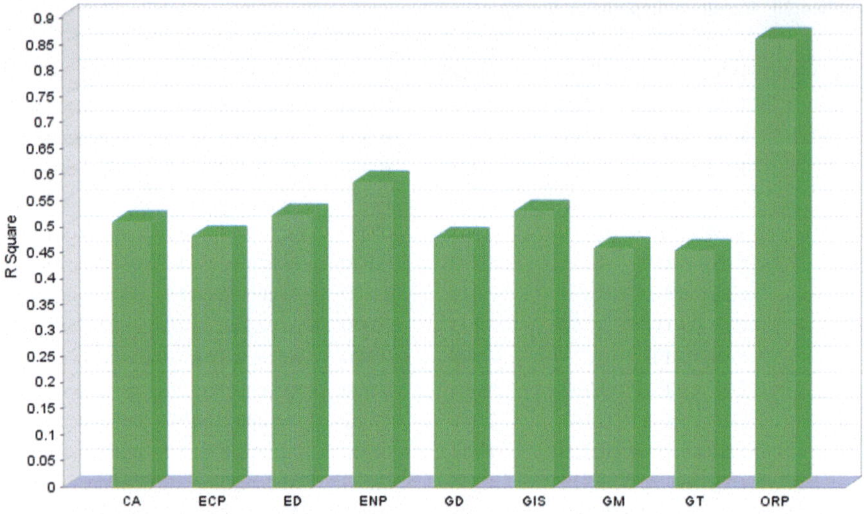

Fig. 4.11 *R*-square graph

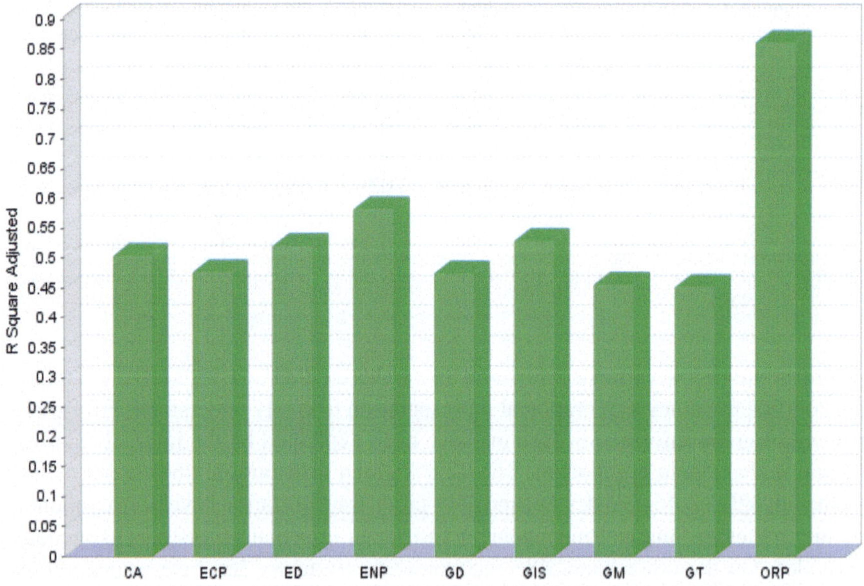

Fig. 4.12 Adjusted *R*-square graph

Table 4.20 indicates the results of hypotheses testing, and we can see the values of *t* statistics and *p* value/sig value for confirmation of structural modeling results. The hypothesis of GT > CA is partially accepted at the 10% confidence level, while the remaining hypothesis is accepted at 1% and 5% confidence level.

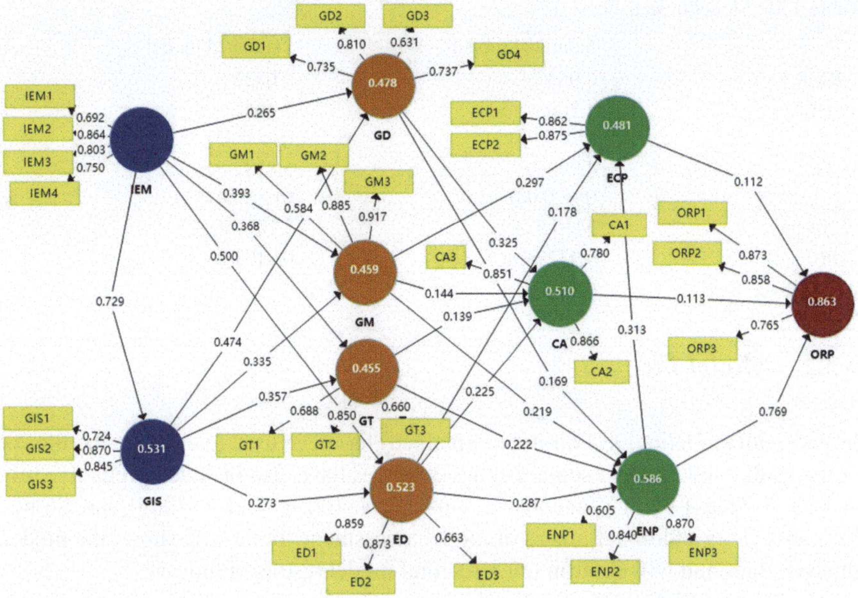

Fig. 4.13 The results of structural model

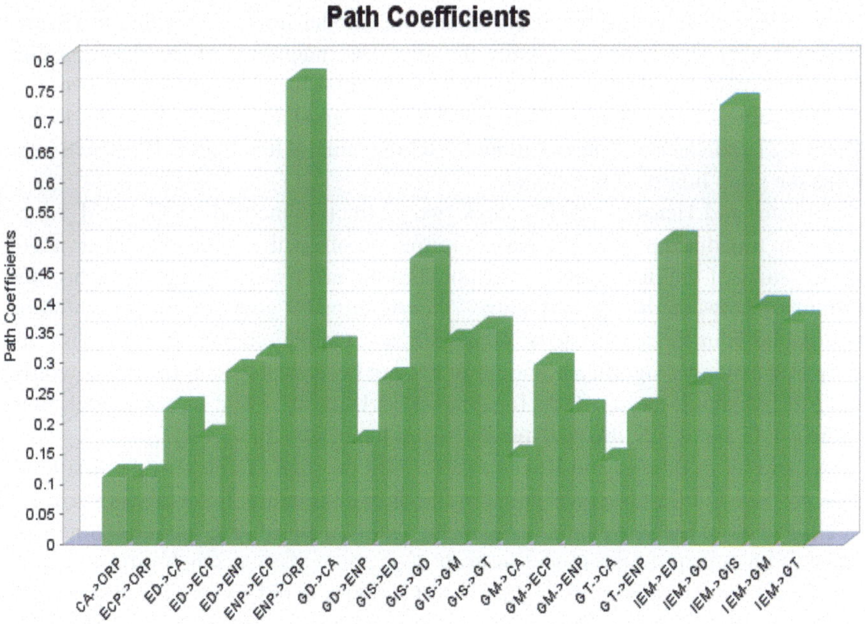

Fig. 4.14 The coefficient of structural model

Table 4.21 Model fit statistics

	Saturated model	Estimated model
SRMR	0.063	0.0822
d_ULS	4.266 ($p > 0.05$)	7.402 ($p > 0.05$)
d_G	4.998 ($p > 0.05$)	5.878 ($p > 0.05$)
Chi-square	5436.574	5736.750
NFI	0.932	0.910
RMS_theta	0.105	

4.5.2 Model Fit

In PLS path modeling, R^2 (variance explained) has been used to evaluate the fitness of the study model, while some recent researches have also included predictive relevance goodness of fit (Tenenhaus, Vinzi, Chatelin, & Lauro, 2005) and Stone-Geisser's Q^2 as additional fitness of model assessment. Table 4.21 shows the model fitness values and will confirm our structural model results in fitness.

The value of SRMR (standardized root mean square residual) is a measure of the mean absolute value of the residuals' covariance, the SRMR based on transforming the predicted covariance matrix and sample covariance matrix into correlation matrices. The standardized root mean square residual is defined as the difference between the correlations of observed and the model implied correlation matrix. However, if it also allows assessing the avg. The magnitude of the discrepancies between expected and observed correlations is an absolute measure fit criterion. The SRMR value of less than 0.10 is considered a good fit (Hu & Bentler, 1999). Table 4.21 indicates that in our model, SRMR value is less than 0.10, which confirms the good fitness of the model.

Dijkstra and Henseler (2015) show two distinctive methods d_G and d_LS to calculate this discrepancy. The bootstrap routine offers the confidence intervals of these values of the discrepancy. Attributed to the error of sampling, therefore, the difference between the implied matrix of correlation by your empirical correlation matrix and your model should be nonsignificant. On the other hand, if the discrepancy is statistically significant, model fit has not been established. In simple words, the p value should be $p > 0.05$ for good model fitness, while in our model, both values, d_G and d_LS, are statistically insignificant (see Table 4.21).

According to Bentler and Bonett (1980), NFI (normal fit index) is the first measure proposed in the structural equation modeling literature. It calculates the value of chi-square of the model and compares it against a significant standard. Since the value of chi-square of the model in itself does not explain as much information to judge the fitness of mode. The value of NFI (normal fit index) between 0 and 1, the near from the 1, the better the fit. Normal fit index values greater than 0.9 represent the good fit and are acceptable (Lohmoller, 1989). In our model, the value of NFI is 0.932, which confirms that the model is excellently fitted (see Table 4.21).

The RMS_theta is the root mean squared residual covariance matrix of the outer model residuals (Lohmoller (1989)). The RMS_theta is very useful to verify purely reflective models, while our model is also based on the reflective concept. The value of RMS_theta should be nearby zero to show good model fitness because it would suggest that the associations among the outer model residuals are minimal. The value of RMS_theta is less than 0.12 which shows a well-fitted model; on the other hand, higher values show a lack (Henseler et al., 2014) of the fitted model. Table 4.21 displays the value of RMS_theta of our model, which is less than 0.12; it means that our model is well-fitted.

4.5.2.1 Additional Model Fit Statistics

The GoF (goodness of fit) measure of the model defined by Tenenhaus, Vinzi, Chatelin, and Lauro (2005) is the geometric mean of both avg. variance extracted and the average of *R*-square of the dependent variables. The objective of the goodness of fit is to account for the study model at both stages, namely, structural and measurement models (Chin, 2010; Henseler & Sarstedt, 2013), with a focus on the overall performance of the model. The GoF formula is as follows:

$$\text{GoF} = \sqrt{(\text{Avg}.R^2 \times \text{Avg.AVE})}$$

GoF shows the goodness of fit.
Avg. R^2 indicates the average of all *R*-square values.
Avg. AVE denotes the average of all the average variance extracted values.

$$\text{GoF} = \sqrt{(\text{Avg}.R^2 \times \text{Avg.AVE})}$$

$$\text{GoF} = \sqrt{(0.543 \times 0.64)}$$

$$\text{GoF} = \sqrt{(0.348}$$

$$\text{GoF} = 0.590)$$

The GoF (goodness of fit) criteria are as follows:

GoF less than 0.1	No fit
GoF between 0.1 and 0.25	Small
GoF between 0.25 and 0.36	Medium
GoF greater than 0.36	Large

In our model, the GoF (goodness of fit) value is 0.590, which is considerable. In other words, our GoF model is large enough to consider sufficient global PLS (partial least square) model validity.

Table 4.22 Predictive relevance (Q^2)

	SSO	SSE	Q^2 (=1–SSE/SSO)
CA	1032.00	690.626	0.331
ECP	688.000	452.264	0.343
ED	1032.00	702.697	0.319
ENP	1032.00	680.205	0.341
GD	1376.00	1053.97	0.234
GIS	1032.00	686.361	0.335
GM	1032.00	741.21	0.282
GT	1032.00	793.129	0.231
ORP	1032.00	453.698	0.560

The predictive relevance Q^2 requires a measure of predictive capability. The recommended approach to examine predictive relevance is called the procedure of blindfolding. The cross-validation test fits soft modeling like a hand in glove (Geisser, 1975; Stone 1974; Wold, 1982). The Q^2 value should be greater than zero, and in our model, Q^2 values are higher than zero, which confirmed the model goodness of fit (see Table 4.22).

4.5.3 *Path Effects*

The findings of the structural model indicate that most of the causal paths are statistically and positively significant. These results are presented in Fig. 4.12. On the other hand, individual latent construct graphical representation is also shown in the following figures.

Figure 4.15 indicates the path coefficient; we can see the positive impact of competitive advantage on organizational performance, while organizational performance is increased by 11.3% by competitive edge factor. On the other hand, Fig. 4.16 displays the path coefficient, and we can see the statistically and significant impact of economic performance on organizational performance. The figure shows that organizational performance can be increased by 11.2% by improvement in firms' competitive edge in the market.

Figure 4.17 shows the path coefficient; we can quickly realize that ecological design has a positive and statistically significant impact on the competitive edge. Besides, a competitive edge can be increased by 22.5% by adopting the green practices of ecological design. On the other hand, Fig. 4.18 indicates the path coefficient of economic performance. The improvement in economic performance increased by 17.8% through adopting ecological design approaches in the manufacturing firms.

Figure 4.19 shows the path coefficient; we can easily see that ecological design has a positive and statistically significant impact on environmental performance. Furthermore, environmental performance can be increased by 28.7% by adopting the green practices of ecological design.

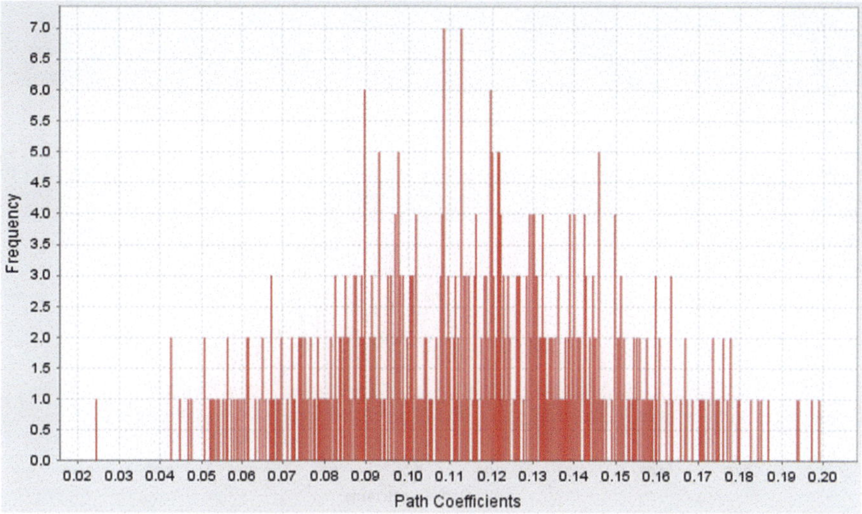

Fig. 4.15 Path coefficient of CA > ORP

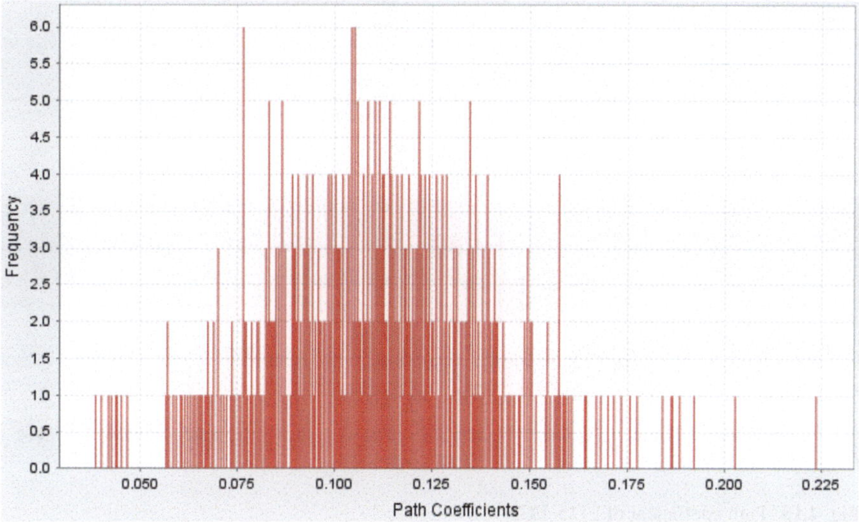

Fig. 4.16 Path coefficient of ECP > ORP

On the other hand, Fig. 4.20 indicates the path coefficient of economic performance. The improvement in economic performance increased by 31.3% through adopting environmental practices in the manufacturing firms of Pakistan.

Figure 4.21 shows the environmental performance impact on the organizational performance, and the results of coefficient path show that organizational

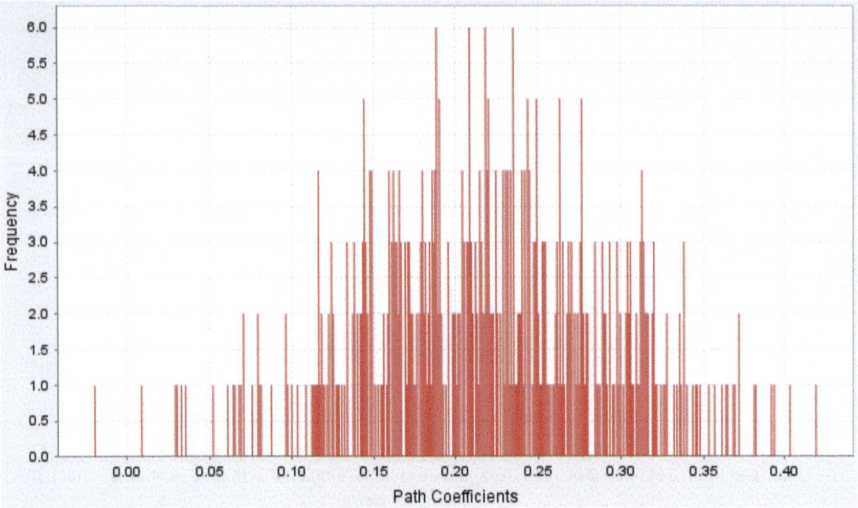

Fig. 4.17 Path coefficient of ED > CA

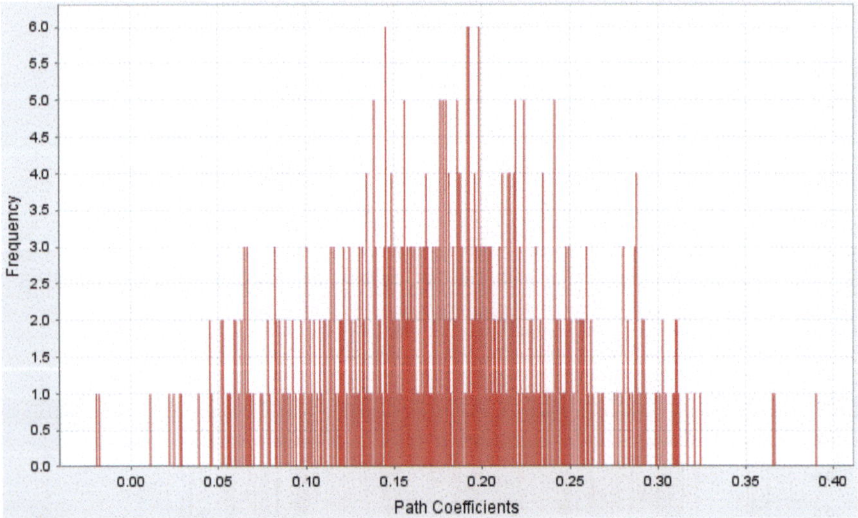

Fig. 4.18 Path coefficient of ED > ECP

performance can be increased by 76.9% through the enhancement in the environmental sustainability and environmental-friendly policy.

Figure 4.22 indicates the green distribution impact on the competitive edge, and the results of the coefficient path show that organizational performance can be increased by 33% through the enhancement in the green distribution systems and implementation of environmental-friendly policy in distribution systems.

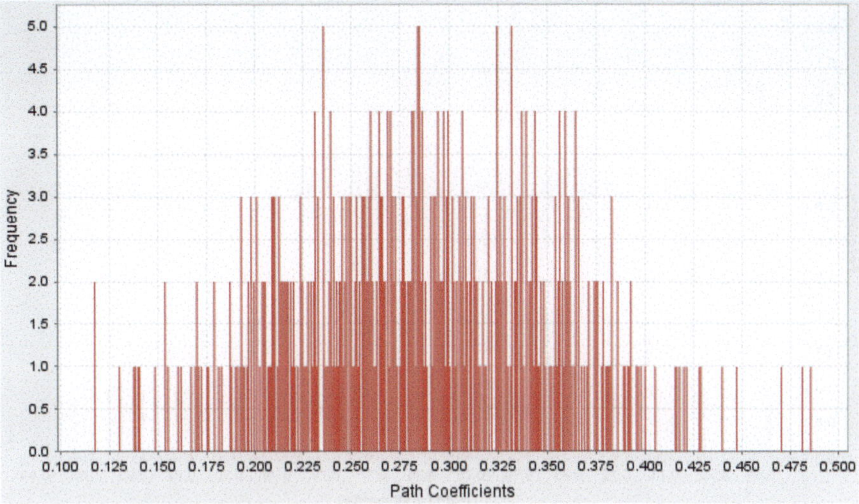

Fig. 4.19 Path coefficient of ED > ENP

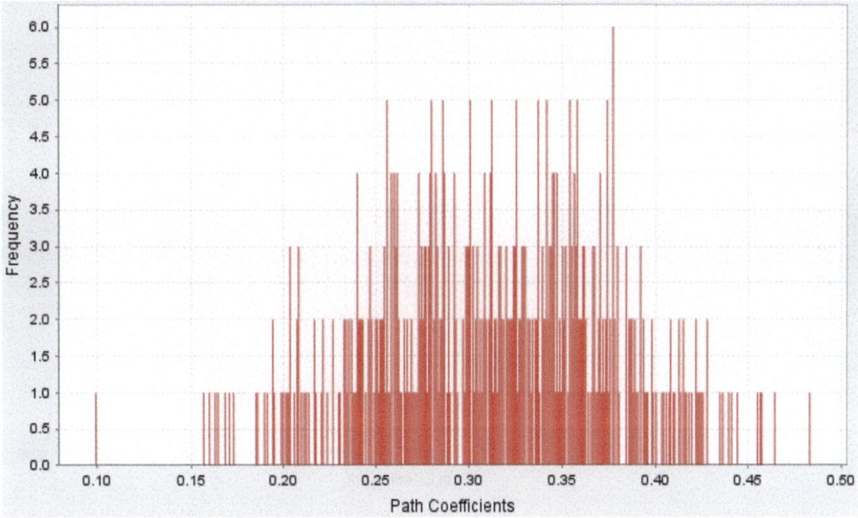

Fig. 4.20 Path coefficient of ENP > ECP

Figure 4.23 indicates the green distribution impact on the environmental performance, and the results of the coefficient path show that environmental performance can be achieved by 16.9% through the enhancement in the green distribution systems and implementation of environmental-friendly policy in distribution systems. On the other hand, ecological design increased by 27.3% by proper implementation of green information systems in end-to-end supply chain management (see Fig. 4.24).

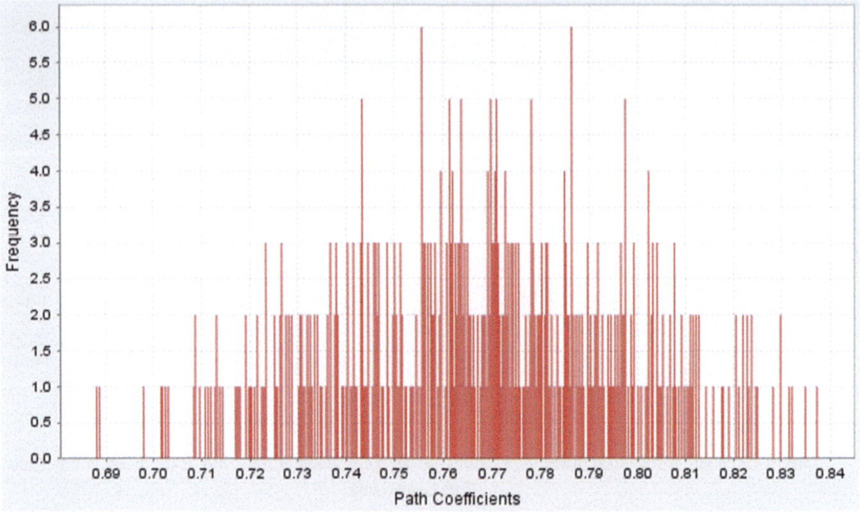

Fig. 4.21 Path coefficient of ENP > ORP

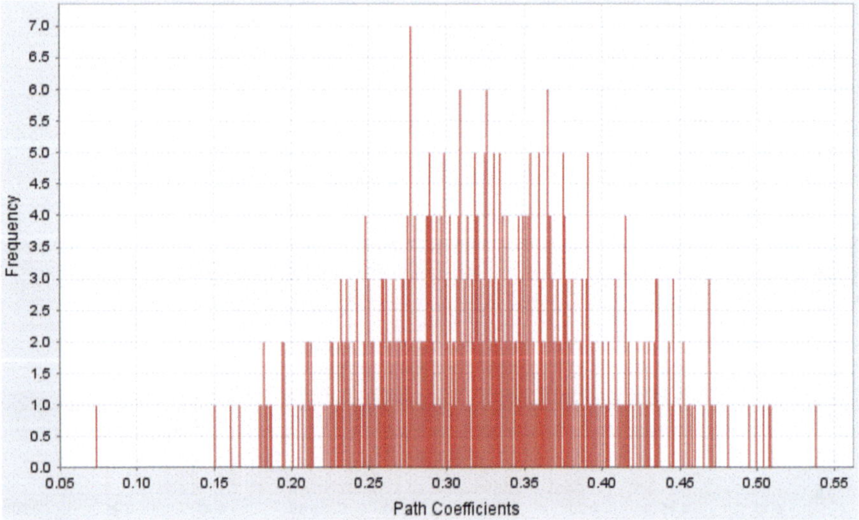

Fig. 4.22 Path coefficient of GD > CA

Figure 4.25 indicates the green information systems impact the green distribution, and the results of the coefficient path revealed that green distribution performance could be increased by 47.4% through adopting green information systems and implementation of the environmental-friendly policy. On the other hand, green manufacturing increased by 33.5% by proper implementation of green information systems in end-to-end supply chain management (see Fig. 4.26).

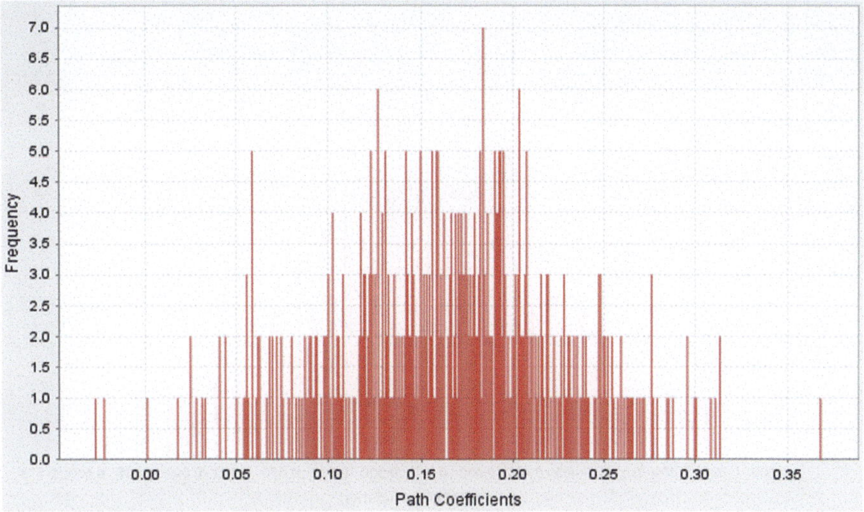

Fig. 4.23 Path coefficient of GD > ENP

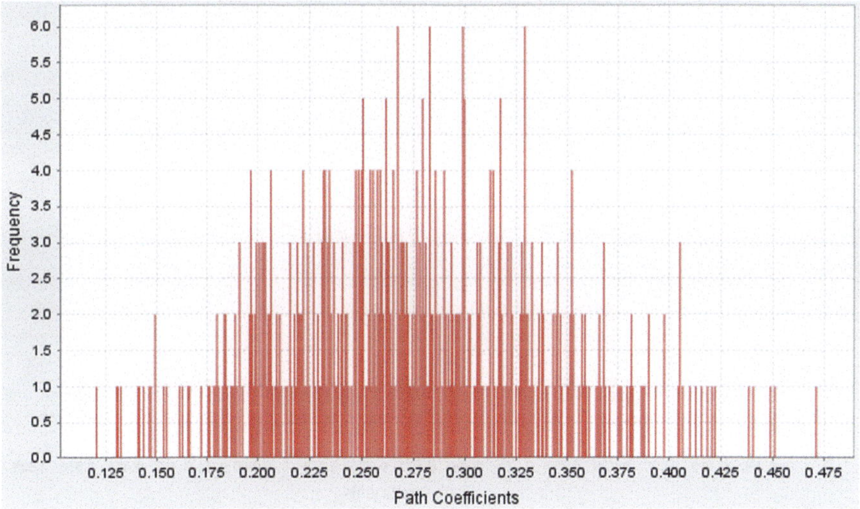

Fig. 4.24 Path coefficient of GIS > ED

Figure 4.27 indicates the green information systems impact the green transportation, and the results of coefficient path revealed that green transportation performance could be increased by 37.7% through adopting green information systems and implementation of the environmental-friendly policy. On the other hand, competitive edge increased 14.4% by proper implementation of green manufacturing systems in end-to-end supply chain management (see Fig. 4.28).

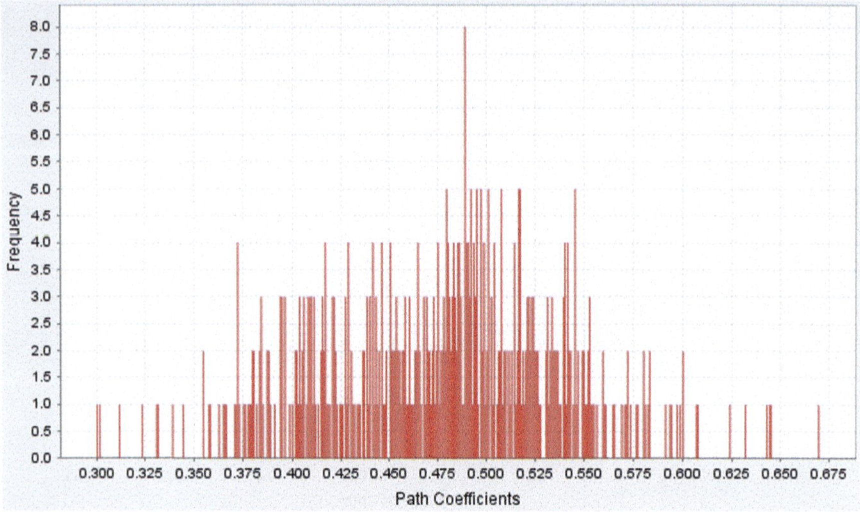

Fig. 4.25 Path coefficient of GIS > GD

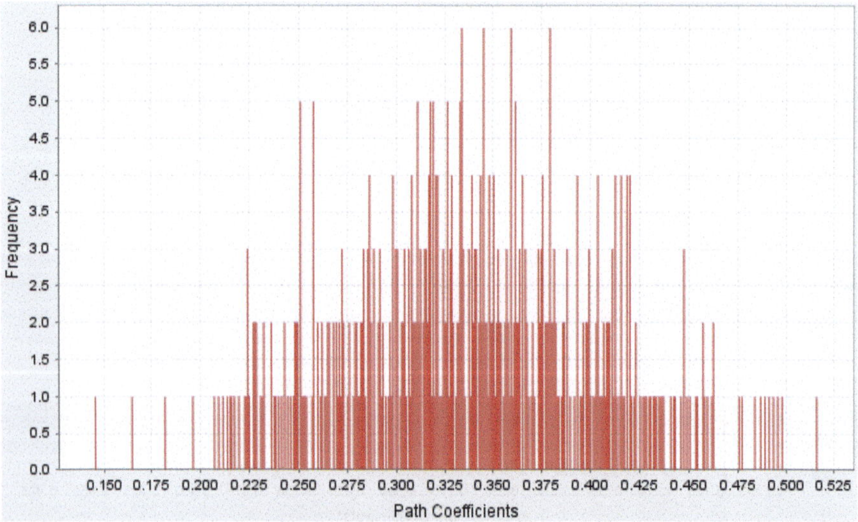

Fig. 4.26 Path coefficient of GIS > GM

Figure 4.29 illustrates the green manufacturing systems' impact on the economic performance, and the findings of the coefficient path revealed that the economic performance of the firms could be achieved by 29.7% through adopting suitable green manufacturing systems and adopting the environmentally sustainable policy framework. On the other hand, environmental performance improved 21.9% by proper implementation of green manufacturing processes in the firms' supply chain management (see Fig. 4.30).

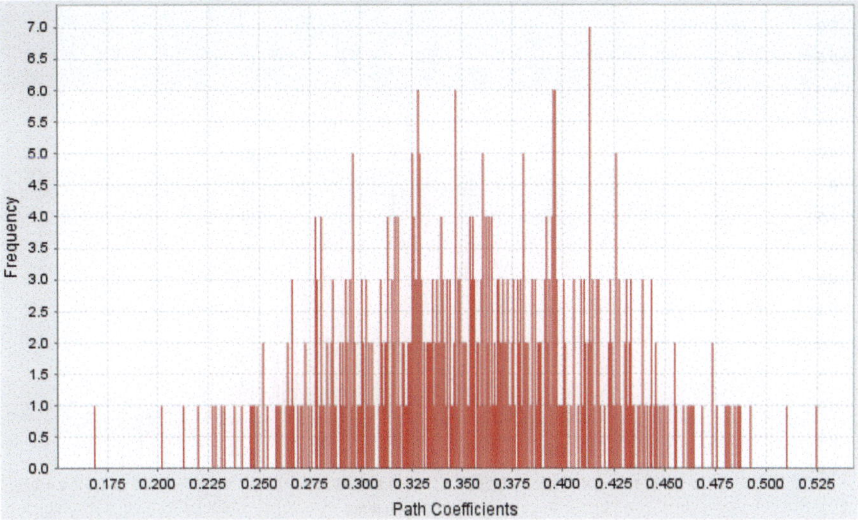

Fig. 4.27 Path coefficient of GIS > GT

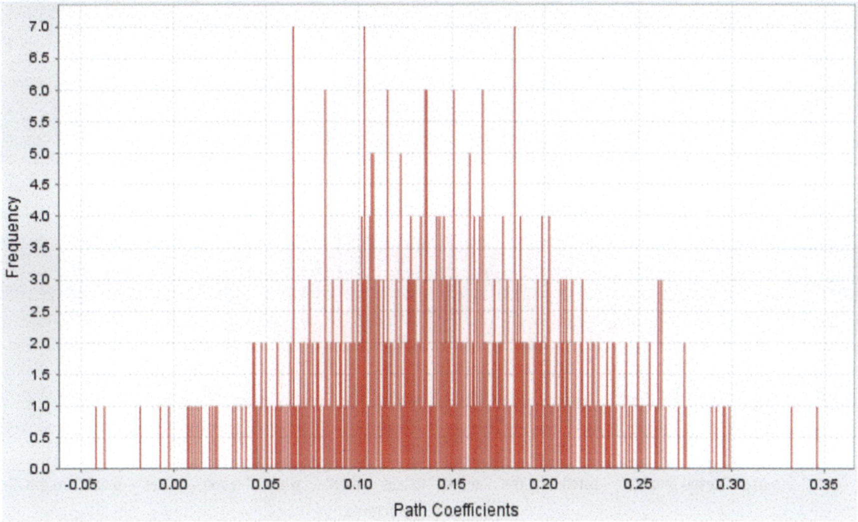

Fig. 4.28 Path coefficient of GM > CA

Figure 4.31 illustrates the green transportation systems' impact on the competitive edge for the firms, and the findings of the coefficient path revealed that the competitive edge of the firms could be achieved by 13.9% through adopting suitable green transportation systems.

Figure 4.32 indicates that the green transportation systems impact the environmental performance of the firms, and the findings of the coefficient path revealed

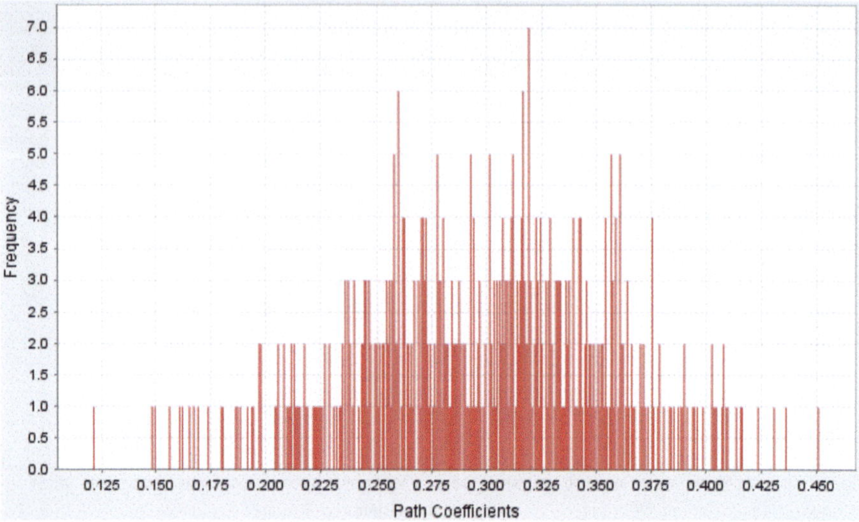

Fig. 4.29 Path coefficient of GM > ECP

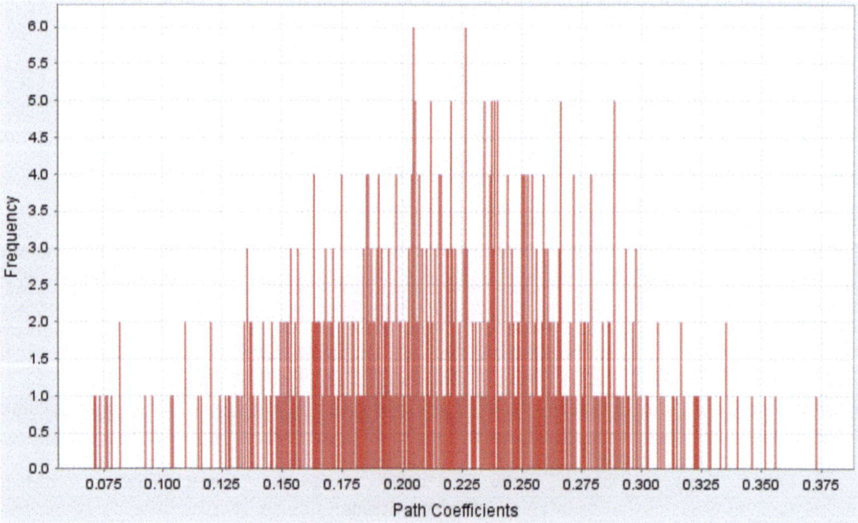

Fig. 4.30 Path coefficient of GM > ENP

that the environmental performance of the firms could be achieved by 22.2% through adopting green transportation systems. On the other side, ecological design can be improved by 50% through internal environmental management systems (Fig. 4.33).

Figure 4.34 shows the internal environmental management's impact on the green distribution of the firms, and the results of the coefficient path show that green distribution of the firms can be achieved by 26.5% through adopting internal

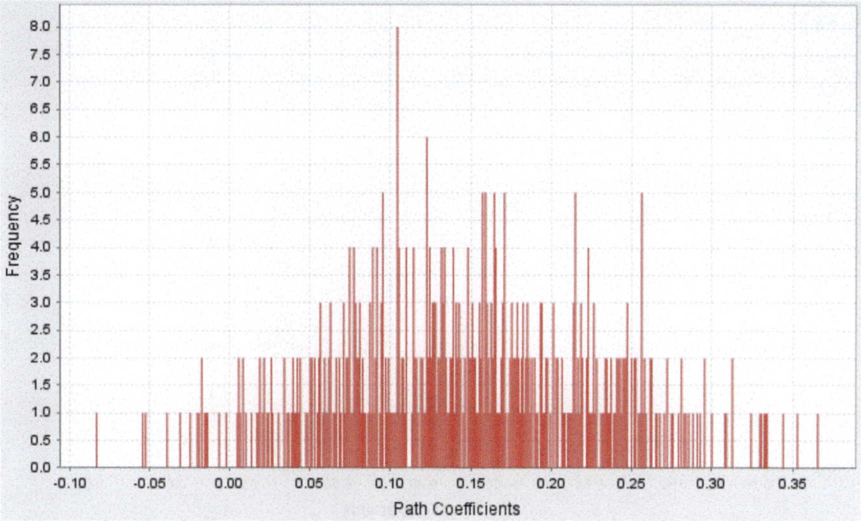

Fig. 4.31 Path coefficient of GT > CA

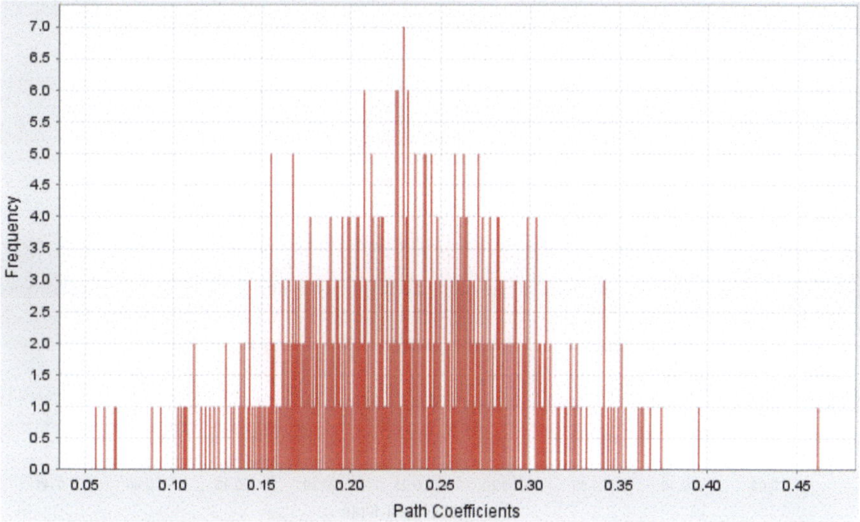

Fig. 4.32 Path coefficient of GT > ENP

environmental management systems. On the other side, green information systems' objective can be achieved by 72.9% through the internal environmental management systems (Fig. 4.35).

Figure 4.36 indicates that the internal environmental management systems impact the green manufacturing of the firms, and the findings of the coefficient path revealed that the environmental-friendly policy manufacturing of the firms

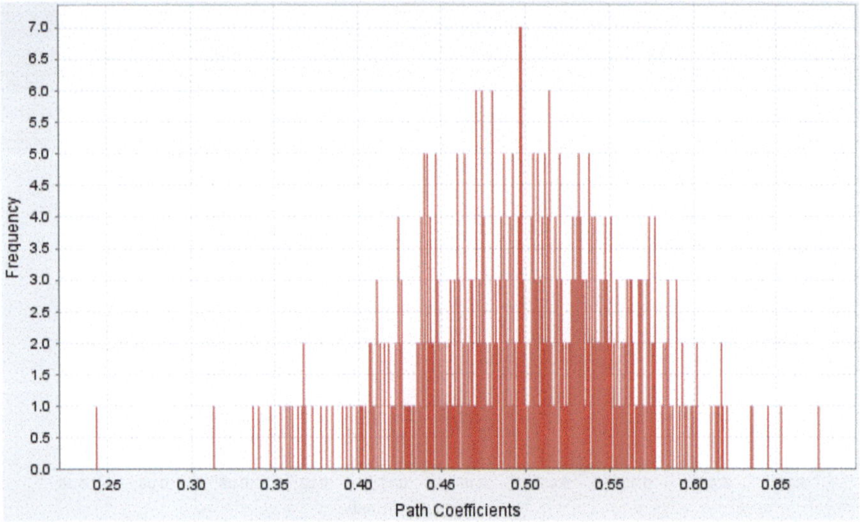

Fig. 4.33 Path coefficient of IEM > ED

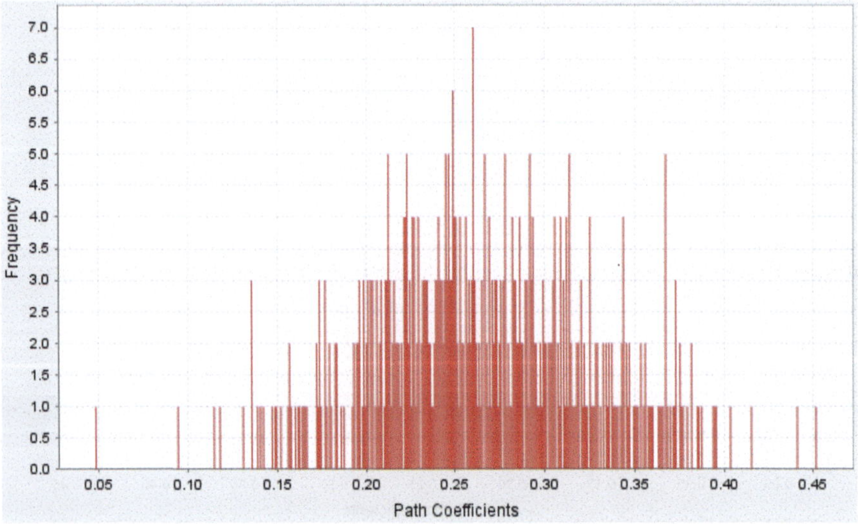

Fig. 4.34 Path coefficient of IEM > GD

could be achieved through the internal environmental management of the firms. Green manufacturing can be achieved by 39.9% through adopting internal environmental management systems. On the other side, green transportation systems can be improved by 36.8% through the internal environmental management systems (see Fig. 4.37).

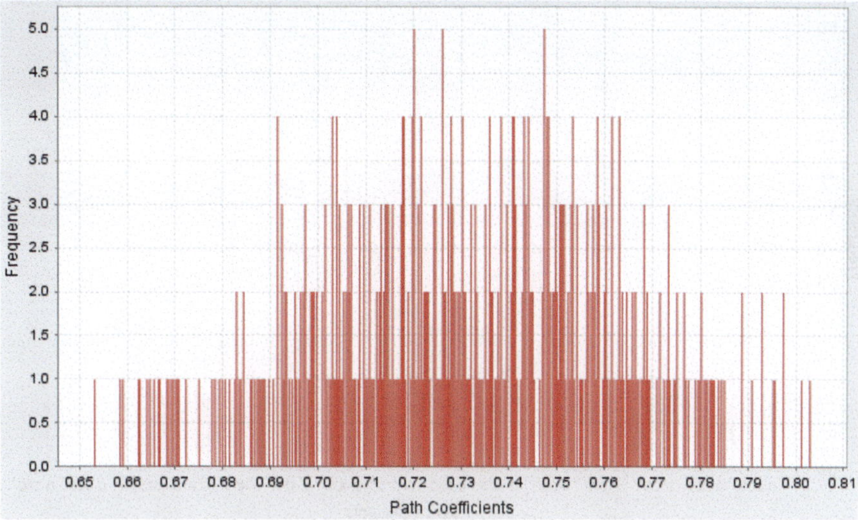

Fig. 4.35 Path coefficient of IEM > GIS

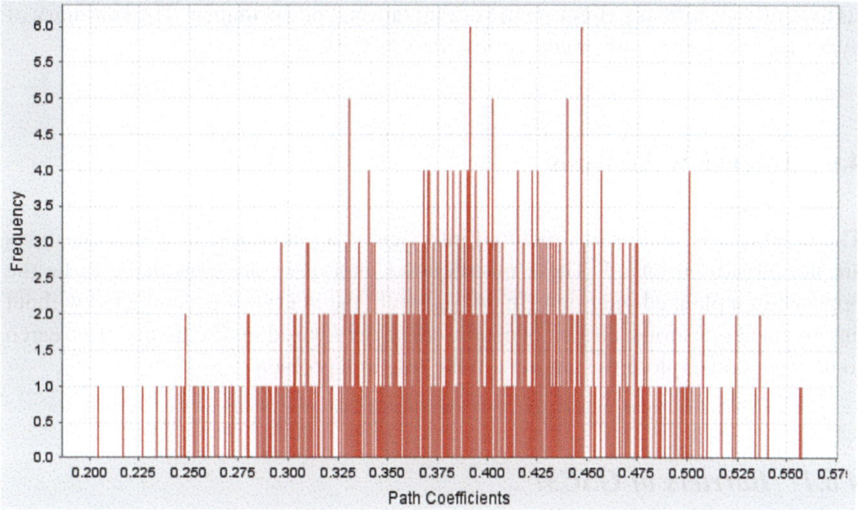

Fig. 4.36 Path coefficient of IEM > GM

4.5.4 Hypothesis Testing

The findings from the model affirm that the green practices, namely, internal environmental management, green information systems, green distribution, green manufacturing, green transportation, ecological design, economic performance, environmental performance, and competitive edge, have a positive and

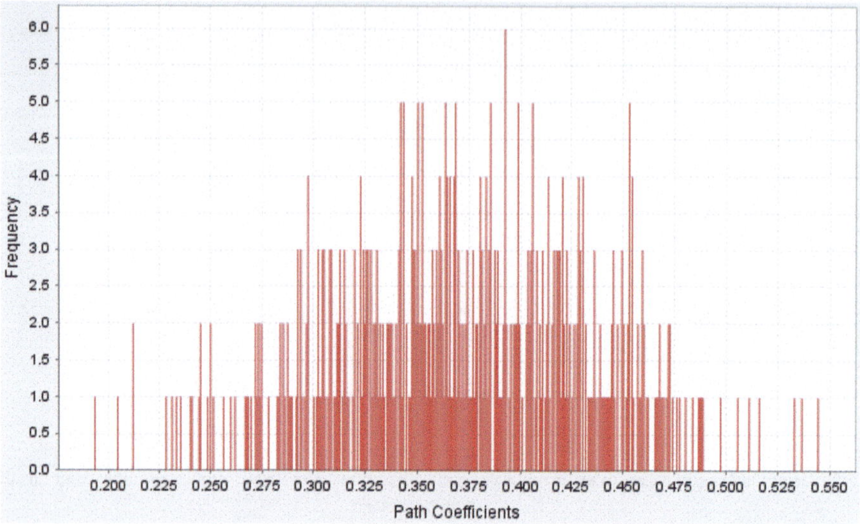

Fig. 4.37 Path coefficient of IEM > GT

statistically significant effect on the organizational performance. The summary of hypothesis testing in this study is provided in Table 4.20.

4.6 Interview Analysis

The nondirective and/or unstructured interview is an interview in which questions are not already arranged. These unstructured interviews are considered to be the opposite of a planned interview. In our research, the interview is conducted without any planned questions, and all questions were only based on the theme of research to uncover and explore the answers to the research problem.

4.6.1 Barriers of GSCM

The barriers of GSCM have been found after the in-depth interview with supply chain professionals. In Table 4.23, barriers have been mentioned in the perspective of manufacturing firms of Pakistan, while in this, some barriers may generally apply to other industries but particularly applicable to the Pakistani manufacturing firms.

The list of barriers is quite long to make it easy; we have divided all the barriers into four parts, namely, economic barrier, environmental barrier, competitive advantage barrier, and organizational barriers. Table 4.24 displays organizational barriers.

Table 4.23 The barriers' faces during the adoption of GSCM practices

S. no.	Barriers
1	Existing poor infrastructure
2	Poor alliance management
3	Lack of middle- and low-level managerial commitment
4	Lack of integration into management
5	Lack of senior management commitment
6	Poor culture and structure of the organization in GSCM
7	Lack of suitable performance management system
8	Poor organizational structure
9	Lack of SC member commitment
10	Firms' policies conflicting with green supply chain management
11	Lack of organizational encouragement
12	Poor quality of labors
13	Lack of knowledge, skills, and experience towards GSCM
14	Lack of understanding among SC members
15	Lack of management initiatives for transport and logistics
16	Lack of interdisciplinary cooperation
17	Differing understanding of term GSCM
18	Lack of functional linkages between indicators – economic, social, and ecological
19	Lack of internal sustainability audits within the firm
20	Lack of environmental-friendly policies (firm level)
21	Complexity in green design
22	Lack of a waste management system or energy management system
23	Lack of experience and training to personnel in GSCM
24	Limited forecasting and planning
25	Resistance to change-lack of trust
26	Lack of skilled human resource professionals in sustainability
27	Old paradigms/culture
28	Unwillingness among SC partners companies and suppliers
29	Lack of employee commitment towards GSCM
30	Lack of ethical concern
31	Lack of human-environment interaction
32	The low ability of one channel member to influence another SC stakeholder
33	Lack of horizontal cooperation between independent firms
34	Unwillingness into the collaborative distribution
35	Lack of public commitment or participation in green projects
36	Multiple complexities and uncertainties
37	Trade barriers
38	Lack of clear definitions of sustainability
39	Pressures from sales
40	Poor conflicting measurement
41	Inconsistent operating goals
42	Security risks

(continued)

Table 4.23 (continued)

S. no.	Barriers
43	Ignorance towards green issues
44	Lack of cooperative behavior of SC stakeholder
45	Lack of research and development of new materials and chemicals
46	Supplier reluctance to change towards green supply chain management
47	Lack of clear understanding of what to solve and how
48	No confidence among customers on green product information
49	Lack of knowledge among customers about ecological labels
50	The low ability of one channel member to control the decision of another
51	Lack of information about the performance of suppliers
52	Lack of robust educational programs on sustainability
53	Lack of ethical concern
54	Lack of human-environment interaction
55	Problems in maintaining environmental suppliers
56	Scarcity of green suppliers
57	Disclosing strategies
58	Customer's unawareness towards GSCM products and services
59	Funding restrictions/inflexibility
60	Inability to incentivize organic providers' cost/budget constraints
61	Estimating hidden costs and potential savings
62	Changing the first-cost mindset
63	Cost of implementation for GSCM
64	The administrative and financial tax burden
65	Financial constraints
66	Massive investments in machinery and infrastructure systems
67	Cost of implementation for GSCM
68	Low financial support from bank side and higher interest rate
69	Lack of financial advantage
70	Lack of long-term profit-oriented goals
71	Capacity constraints
72	Lack of direct incentives
73	Pressures from laws and regulations
74	Absence of take-back legislation
75	Lack of government support for green projects
76	Poor legislation
77	Lack of government subsidy
78	Lack of strict environmental laws
79	Lack of value-added technology
80	Lack of IT consultants and experts
81	Fear of failure
82	Lack of integration of technology/compatibility
83	Poor IT human resource
84	The complexity of measuring/monitoring green practices of suppliers
85	Substantial investment in information technology and software

Table 4.24 The organizational barriers

S. no.	Organizational barriers
1	Lack of middle- and low-level managerial commitment
2	Lack of integration into management
3	Existing poor infrastructure
4	Poor alliance management
5	Poor culture and structure of the organization in GSCM
6	Lack of suitable performance management system
7	Lack of SC members' commitment
8	Lack of internal sustainability audits within the firm
9	Poor organizational structure
10	Resistance to change-lack of trust
11	Lack of senior management commitment
12	Lack of environmental-friendly policies (firm level)
13	Lack of a waste management system or energy management system
14	Lack of organizational encouragement
15	Limited forecasting and planning
16	Lack of understanding among SC members
17	Lack of employee commitment towards GSCM
18	Lack of functional linkages between indicators – economic, social, and ecological
19	Lack of human-environment interaction
20	Lack of management initiatives for transport and logistics
21	The low ability of one channel member to influence another SC stakeholder
22	Inconsistent operating goals
23	Lack of ethical concern
24	Security risks
25	Lack of horizontal cooperation between independent firms
26	Unwillingness into the collaborative distribution
27	Ignorance towards green issues
28	Lack of cooperative behavior of SC stakeholder
29	Lack of research and development of new materials and chemicals
30	Lack of clear understanding of what to solve and how
31	The low ability of one channel member to control the decision of another
32	Lack of information about the performance of suppliers
33	Lack of robust educational programs on sustainability
34	Disclosing strategies
35	Capacity constraints
36	Pressures from laws and regulations
37	Lack of value-added technology
38	Fear of failure
39	Lack of integration of technology/compatibility
40	Poor IT human resource

4.6.1.1 Organizational Barriers

Table 4.24 shows the organizational barriers; it is a long list of barriers; we will discuss a couple of critical barriers in detail such as lack of top and middle management support, lack of stakeholder commitment, firms' policies conflicting with GSCM practices, lack of knowledge and skills, lack of internal sustainability audits and monitoring, and lack of training provided to firms' employees.

Lack of Management Support

The lack of senior management support is identified as very significantly important barriers during the adoption of green supply chain practices. In the market of B2B, companies' green supply chain management activities are not constrained by strategic essentials, and hence, senior management commitment is critically important, while in this market, customers' demand risk of negative media attention forces the execution of green supply chain management activities. Therefore, the senior management commitment and customer demand act as significant barriers to green supply chain management.

Existing Poor Infrastructure

The poor infrastructure of organizations becomes a barrier during the adoption of green supply chain practices. Usually, firms want to implement green supply chains such as green manufacturing, green distribution, green transportation, and green warehousing, but the organizational structure is not able to support green practices. That is why the infrastructure of the organization has a crucial barrier in Pakistani manufacturing firms.

Lack of Understanding Among SC Members

The lack of green supply chain and green project understanding among supply chain partners is a significant and severe barrier because most times, firms started green projects to compete with their competitors without realizing and coordinating with their supply chain members, and after all, firms bear the high cost in terms of the failure of projects and also built a negative image in customers.

Lack of Employee Commitment Towards GSCM

Employees are the internal power of firms, and employees also represent their firms in public. If employees are not committed to the adoption of green supply chain management practices, it means that firms have no power and ability to convince their employees to be dedicated to the implementation of green projects. On the other hand, firms cannot successfully implement green projects without the involvement of their employees. Besides, customers also cannot be satisfied with firms' practices because ultimately, employees are the representative of the firms.

Lack of Ethical Concern

Behind the adoption of green supply chain practices and sustainability practices are ethical values of firms. If firms are lacking in ethical concerns, it is almost impossible to go towards green projects and environmental-friendly policy supply chain practice implementations. Besides, lack of ethical concern is a crucial barrier faced by Pakistani manufacturing firms. On the other hand, firms have strong ethical values and easily and comfortably adopt green practices in their end-to-end supply chain processes without any resistance of employees and supply chain stakeholders/members.

Lack of Clear Understanding of What to Solve and hHow

The lack of clear understanding is a significant barrier to green supply chain practice implementations. Because most times, firms do not know what to solve and how to solve, means which are the most polluting operations in the firms need to change/improved and how to start green projects. These are very important and create a substantial effect on firms' organizational strategy. On the other hand, in the manufacturing firms of Pakistan, many firms are facing the problems of lack of green supply chain understanding.

Lack of Information About the Performance of Suppliers

In the perspective of Pakistani manufacturing firms, most times, companies are unable to get/receive the green performance information of their first-tier and second-tier suppliers, which not only discourage firms from going towards green practices but firms' customers also raise questions. Furthermore, suppliers are not willing to share their green practices with manufacturers due to a lack of trust and fair of loose competitive advantage.

Lack of Robust Educational Programs on Sustainability

In successful green projects, companies educate their employees through sustainability programs and benefits. Most firms conducted sustainability programs on a weekly and monthly basis. Undeniably, lack of robust education programs on sustainability does not obsolete the knowledge of employees but also loses their confidence in their involvement in green projects.

Fear of Failure

Fear of failure is the main obstacle for firms, which stops to go towards green and sustainability programs/projects. Fear of failure has been divided into different categories, such as financial, image and reputation, supply chain partners, etc. In the context of Pakistani manufacturing firms, almost 30% of firms are falling in different fear of failure, which mainly are financial constraints and scarcity of experts.

Poor IT Human Resource

Firms' driving forces are capital, human, and resources. There is no doubt that human resource is critical factors in the success of firms. The green supply chain projects required highly qualified consultants and experts, which can integrate all the systems. From the perspective of Pakistan, manufacturing firms are facing a scarcity of skilled and professional workforce in terms of IT consultants and experts, which not only resist firms to change their mindset and not to adopt green practices and avoid risk.

4.6.1.2 Economic/Financial Barriers

The financial barriers cover initial investment in green information technology and slow ROI (return on investment) after the execution of the green project. Mostly green machinery, materials, and processes to control environmental pollution are costly. The firms are unable to implement green projects due to financial crises or limitations of the financial budget. On the other hand, green project return is also slow as compared to non-green projects due to lack of green practice understanding, employee training, scarcity of professional IT experts, etc. (see Table 4.25).

Cost of Implementation

The cost of a green supply chain management project is higher than non-green supply chain projects. Green projects are required to change sometime organizational strategy and/or supply chain strategy. Due to green ideology, firms need to accept change management, change attitude, and change behavior with customers such as firms need to select green suppliers for buying green materials, require green distributors/retailers, and need to be closer with customers/end users to fulfill their requirement and brief them about green products. These steps incur a considerable cost in the supply chain system and increase the financial burden on the firms' profitability.

Table 4.25 The economic barriers

S. no.	Economic barriers
1	Trade barriers
2	Cost of implementation for GSCM
3	Funding restrictions/inflexibility
4	Inconsistent operating goals
5	Lack of research and development of new materials and chemicals
6	Poor conflicting measurement
7	Supplier reluctance to change towards green supply chain management
8	Pressure from sales
9	No confidence among customers on green product information
10	Lack of knowledge among customers about ecological labels
11	Customers' unawareness towards GSCM products and services
12	Inability to incentivize organic providers' cost/budget constraints
13	Estimating hidden costs and potential savings
14	Changing the first-cost mindset
15	The administrative and financial tax burden
16	Financial constraints
17	Significant investments in machinery and infrastructure systems
18	Low financial support from bank side or higher interest rate
19	Lack of financial advantage
20	Lack of long-term profit-oriented goals
21	Lack of direct incentives
22	Lack of government financial support for green projects
23	Lack of government subsidy
24	Massive investment in information technology and software

Lack of Direct Incentives

The lack of direct incentives discourages firms from adopting green practices in their supply chain management. The green supply chain practices cannot bring overnight change in a firm's economic performance and operational performance. This is a concept of green to protect environmental sustainability; the primary purpose of the green supply chain is to make the environment safe, reduce carbon emissions, and reduce the usage of hazardous chemicals in the products, while increasing the profit is the sub-objective of the green supply chain. Many firms implement green practices in their business activities to increase the firm's performance, which is the wrong perception; substantial financial gain is not the primary objective of the GSCM. On the other hand, the financial gain cannot achieve dramatically; it also takes time, and usually, it depends on the products manufactured by firms and their relationship with customers, which means how they deliver the right message of green products at the right time to the customers.

Higher Interest Rate on the Loan

The higher interest rate is a significant factor and plays a vital role in discouraging the firms. Because there are no such direct benefits received by banks due to the implementation of green projects, in European countries, banks offer the lowest interest rate to the firms for the adoption of green projects and encourage firms to implement green practices in their firms. However, from the perspective of Pakistan, still, there is no such direct incentive provided by government/private banks.

Trade Barriers

The barriers of trade play a vital role in the context of Pakistan. Because some manufacturers have their customers in African countries, they do not need green products but cheap products. Also, most African countries are working on green products but still in the very narrow market and limited product lines. Furthermore, if company manufacturer green products and dispatch to African customers so their competitor manufacturer same non-green products at a cheap cost and sell to the customers, which not only grab to the African market but also ear healthy profit.

Lack of Government Subsidy

The lack of government support most times plays a significant role during the adoption of green supply chain practices. The government is not providing as such moral and financial support to the firms, while firms' competitors are producing non-green products at the lowest cost. In the context of Pakistan, manufacturing firms are facing a lack of government subsidy support because the government is not providing as such handsome financial support in terms of subsidy or tax exemptions for green projects.

Heavy Investment in Information Technology and Software

The adoption of green supply chain practices requires substantial investment in terms of information technology and software compatibility with suppliers and distributors for information sharing and collaboration. From the perspective of Pakistan, manufacturing firms are mainly facing this issue and avoiding/ignoring the needs of green products. However, in the real case, they are also losing the opportunity to grab new markets and customers.

4.6.1.3 Environmental Barriers

The environmental barriers cover firms' policies conflicting with green supply chain management; complexity in green design; lack of environmental-friendly policies (firm level); lack of knowledge, skills, and experience towards GSCM; lack of experience and training to personnel in GSCM; lack of human-environment interaction; lack of government support for green projects; lack of clear definitions of sustainability; and lack of strict environmental laws. The complete list of economic barriers is mentioned in Table 4.26.

Firms' Policies Conflicting with Green Supply Chain Management

Many times, firms' policies are conflicting with green supply chain management practices. Companies would like to adopt green supply chain operations in their manufacturing and logistics operations. However, firms' policies are rigid and not flexible to mold them into greener processes. Besides, in the context of Pakistan, several enterprises want to implement green supply chain practices in their firms for gaining competitive advantage and economic and environmental benefits, but because of rigid policies, they are unable to adopt green practices quickly. In simple words, firms' policies conflicting with green supply chain management practices are the key barrier found in manufacturing firms of Pakistan.

Complexity in Green Design

The green supply chain management concept is mainly focused on the ecological design of products. Green products reduce the usage of harmful chemicals and materials, but green products also mitigate environmental degradation with smooth and comfortable recycling, remanufacturing, and refilling techniques. In the context of Pakistan, we found that firms are facing "complexity in green design." Most firms have no experience "how to convert traditional supply chain to a green supply chain." Furthermore, the scarcity of professional workers and consultants is also playing a vital role in ignoring green supply chain techniques.

Lack of Environmental-Friendly Policies

The lack of environmental-friendly policies is reflecting as a barrier to the adoption of green supply chain management. Many times, companies are ignoring and avoiding to adopt green supply chain practices in their firms due to a lack of support from the government side in terms of lack of environmental-friendly policies. Besides, the lack of environmental-friendly policies also encourages firms to continue production with "polluted supply chain/traditional supply chain techniques."

Table 4.26 The environmental barriers

S. no.	Environmental barriers
1	Firms' policies conflicting with green supply chain management
2	Lack of understanding among SC members
3	Lack of research and development of new materials and chemicals
4	Lack of functional linkages between indicators – economic, social, and ecological
5	Complexity in green design
6	Lack of a waste management system or energy management system
7	Lack of management initiatives for transport and logistics
8	Lack of internal sustainability audits within the firm
9	Lack of environmental-friendly policies (firm level)
10	Lack of knowledge, skills, and experience towards GSCM
11	Lack of employee commitment towards GSCM
12	Lack of experience and training to personnel in GSCM
13	Lack of ethical concern
14	Lack of human-environment interaction
15	Lack of skilled human resource professionals in sustainability
16	Old paradigms/culture
17	Lack of public commitment or participation in green projects
18	Lack of clear definitions of sustainability
19	Ignorance towards green issues
20	Supplier reluctance to change towards green supply chain management
21	Lack of clear understanding of what to solve and how
22	Lack of information about the performance of suppliers
23	Lack of robust educational programs on sustainability
24	Problems in maintaining environmental suppliers
25	Scarcity of green suppliers
26	Pressures from laws and regulations
27	Lack of government support for green projects
28	The complexity of measuring/monitoring green practices of suppliers
29	Lack of strict environmental laws
30	Poor legislation

Lack of Knowledge, Skills, and Experience Towards GSCM

The lack of green supply chain knowledge, experience, and skills is the significant barrier faced by manufacturing firms in Pakistan. However, they are unable to find an experienced consultant, and it became a key barrier for firms. Furthermore, scarcity of IT and supply chain consultants is also a problem faced by manufacturing firms. In fact, due to a limited number of experts available in Pakistan, it is costly to hire consultants for green projects.

Lack of Clear Definitions of Sustainability

The lack of a clear definition of green supply chain management/sustainability becomes a barrier in the manufacturing firms of Pakistan. Because of the scarcity of supply chain consultants/experts, each firm defined green supply chain/sustainability definition itself, which is different from their supply chain partners including suppliers and distributors, and finally, it creates a different situation for firms. As we found, several firms adopt green supply chain techniques only for gaining competitive advantage and economic benefits. On the other hand, several firms defined green supply chain definition as to protecting environmental sustainability and not counting economic benefits.

Lack of Government Support for Green Projects

During the interview with firms, the lack of government support for green projects is found as a potential barrier. Undeniably, several new environmental laws have been developed by the federal and provincial governments of Pakistan. Nevertheless, still, there is more improvement required in terms of financial and physical support such as provide loans on the lowest interest rate, tax exemptions on green projects and green supply chain firms, government subsidies, and encouragement through annual appraisal/inspections of manufacturing firms' practices.

Lack of Experience and Training to Personnel in GSCM

The lack of experience of GSCM projects mainly can also be boosted through the governmental level to provide in-detailed financial and environmental benefits to firms' senior management. Also, in some European countries, regulatory authorities provide free courses/lectures on green supply chain techniques and methodologies, which give an idea to senior management and also encouragement. In the context of Pakistan, the government is not providing such training to firms.

Lack of Strict Environmental Laws

The strict environmental laws create pressure on firms to adopt green and environmental-friendly policy practices in their end-to-end supply chain management operations. Undeniably, in the last couple of years, the Pakistani government has passed some new environmental laws, which is very beneficial, and firms are eager to adopt green supply chain practices in their firms. Besides, the UN (United Nations) also appreciates the efforts of the Pakistani government in terms of "Billion Tree Project" and "Sustainability" projects. However, still, further, the improvement required to boost environmental sustainability to a large extent.

4.6.1.4 Competitive Edge Barriers

Many competitive edge barriers are found during an interview with senior managers of manufacturing firms of Pakistan. The barriers included in competitive edge are poor quality of labor, no confidence among customers on green product information, lack of skilled human resource professionals in sustainability, lack of integration of technology/compatibility, poor IT human resource, and lack of IT consultants and experts. The complete list of economic barriers is mentioned in Table 4.27.

Lack of IT Consultants and Experts

The lack of IT experts and consultants is a significant barrier found in the manufacturing firms of Pakistan. There is no doubt that IT consultants and experts provide a critical competitive advantage to the firms, and firms are willing to pay a higher salary and benefits to the IT consultants. However, unfortunately, in Pakistan, IT consultants and experts are very few to integrate overall green supply chain systems with suppliers and distributors for greater information sharing.

No Confidence Among Customers on Green Product Information

Usually, firms gain benefits and create a competitive advantage when customers trust firms' green products and green labels on products. However, in Pakistan, one segment of customers does not believe in firms' green products, and according to them, firms are still producing polluted products, but they labeled them green. Undeniably, this is a barrier for other firms who are willing to adopt green supply chain practices for competitive advantage but whenever, they realize this issue so avoid to invest huge money in green projects and also avoid to take risks. On the other hand, a large segment of customers has started a trust in firms' green practices due to annual inspections of federal and provincial governments.

Poor Quality of Labors

The poor quality of labors is faced by Pakistani manufacturing firms, and much senior management identifies this factor as a barrier. According to the firms, they are almost unable to find highly qualified and experienced supply chain experts in small cities, which is not only creating problems for firms but also for the environment. Poor quality of labor is also a risk for firm reputation and image.

Table 4.27 The competitive edge barriers

S. no.	Competitive edge barriers
1	Lack of knowledge, skills, and experience towards GSCM
2	Lack of SC members' commitment
3	Existing poor infrastructure
4	Poor alliance management
5	Lack of suitable performance management system
6	Poor quality of labors
7	Limited forecasting and planning
8	Old paradigms/culture
9	Lack of understanding among SC members
10	Lack of employee commitment towards GSCM
11	Lack of functional linkages between indicators – economic, social, and ecological
12	No confidence among customers on green product information
13	Lack of experience and training to personnel in GSCM
14	Lack of knowledge among customers about ecological labels
15	Lack of skilled human resource professionals in sustainability
16	Lack of value-added technology
17	Lack of IT consultants and experts
18	Lack of integration of technology/compatibility
19	Poor IT human resource

Lack of Integration of Technology/Compatibility

The lack of integration with suppliers and distributors systems, mostly times suppliers and distributors, are not agreed to share their information with manufacturing due to confidentiality/privacy reasons. Also, the compatibility of technology is a problem that exists in the integration of suppliers' systems with distributors and manufacturers. Finally, due to the lack of integration/compatibility, information sharing is not trustable/reliable, which is to create serious risks.

Poor IT Human Resource/Lack of Skilled Professionals for Green Projects

This is the most significant risk for firms but also for environmental sustainability. Many firms are unable to adopt green supply chain practices properly due to fear of failure. In addition, scarcity of highly qualified green project experts' one side does not create a competitive advantage for firms. However, on the other hand, it is a risk for firms to higher unskilled professionals for their green supply chain project implementation. Furthermore, highly qualified green supply chain experts significantly provide a competitive advantage for firms.

4.6.2 *Drivers of GSCM*

The green supply chain management has several drivers and/or motivating factors, which influence firms to implement complete or partial green practices in their end-to-end supply chain system. Broadly the drivers of the green supply chain (GSCM) are four types: (1) Economic performance drivers, (2) Environmental performance drivers, (3) Competitive advantage drivers, and (4) Organizational performance drivers. These drivers and factors influence to adopt green practices, while as we have done our research in the perspective of Pakistani manufacturing firms, we have found many drivers which may be generally implemented on every industry and some drivers are specified or attached with the manufacturing firms in Pakistan, due to governmental policies, cultural factors, organizational structure, customer awareness, management commitment, etc. Table 4.28 displays all drivers and motivating factors of green supply chain management, including economic performance drivers, environmental performance drivers, competitive advantage drivers, and organizational performance drivers.

4.6.2.1 Economic Performance Motivating Factors

The economic performance drivers or motivating factors include recycle within the firm, lower rejection and defective products, compliance with sectorial price behavior transportation cost and production cost, government subsidies and incentives, new market opportunities, reusable packages, lower consumption of materials and packaging, increase market share, and increase resources with limited waste. Table 4.29 shows the several economic performance drivers and/or motivating factors of GSCM (green supply chain management). Due to the long list of economic performance drivers, we will discuss some of the motivating factors in detail to build an understanding of how these factors play a motivating force in the GSCM.

Special Tax Exemption for ISO-Certified Firms

The tax exemption is provided by regulatory authority/government to the green projects or green practices implemented in firms. Undeniably, tax exemption is playing a supporting role in encouraging firms to reduce carbon emissions, GHG (greenhouse gases), and global warming and control climate change through implemented green supply chain management.

Greater Export Opportunities

Firms also implement green practices in their manufacturing supply chain to increase their market share and grab export opportunities in Europe. Because many European and American firms are not willing to purchase polluted materials from suppliers or custom regulatory authority, they do not allow firms to import polluted products. Ultimately, green products have more export opportunities as compared to non-green products. In the perspective of Pakistan, the government also has environmental laws to be followed by import and export firms, while in the last couple of years, environmental policies are more strict due to international pressure on federal and provincial governments.

Recycle Within the Firm

In Pakistan, many companies are using the "recycling within-firm" technique to increase their economic performance. Undeniably, recycle within firms provide several benefits mainly including recycling business waste means less landfill and less disposal and less overall harm to the environment; firm can attract new customers, improve chances of winning contracts, and enhance customer loyalty by demonstrating environmental responsibility through recycling efforts; managing and handling waste is costly, and reducing the among of waste firm send directly to landfill can bring substantial savings on landfill tax; businesses in some industries have a legal responsibility for disposal of specific products; and ensuring compliance through recycling schemes means avoiding penalties and fines.

Lower Rejection and Defective Products

Firms save handsome money in terms of lower rejection rates and defective products. In Pakistan, firms are adopting green practices to improve their processes and reduce their end-to-end system waste.

Internal Reduction of Production Costs

In Pakistan, firms are implementing green supply chain practices in their manufacturing processes due to the reduction in costs. Lowering your production costs can be one of the best ways to increase revenue and make your firm more profitable. Firms use green practices to reduce their production costs. This method of manufacturing aims to cut out the "fat," or waste, associated with the manufacturing processes. In this case, waste is defined as anything that does not add value to the customer.

Government Subsidies and Incentives

Several firms were adopting environmental-friendly policy practices for improving their economic performance in terms of government subsidies and incentives, etc. On the other hand, the government is usually offering financial incentives, such as tax breaks and subsidies, to select firms. However, financial incentives used to entice businesses to come to the taxpayers' expenses. Besides, manufacturing firms are also getting funds from international organizations, business unions, and nongovernment organizations in terms of appreciation and rewards.

Early Supplier Involvement in Green Product Development

Early involvement of suppliers in green product development offers substantial economic benefits because most of the time, suppliers have more broad experiences than the manufacturer, which helps the manufacturer reduce the cost of green products and processes. Besides, early supplier involvement in green projects also offers "previous green projects experienced," which is significantly beneficial for the firms. In the context of Pakistan, firms are involving their suppliers in the initial stage of product for maximum "experience sharing." Involving suppliers in cross-functional teams at the early stages of product development has strong roots in the Pakistani manufacturing industry. Today early supplier involvement remains quite common in manufacturing and consumer electronics industries. Many purchasing organizations view coordination with critical suppliers via ESI as essential enablers to the product, process, and supply chain structure development and as a cost reduction exercise. Besides, adopting ESI practices may offer additional benefits to organizations, including the management of supply risk in new product development and the upstream supply chain. For a supplier, participation may be embedded in the already existing partnership or alliance with the manufacturer or a way of securing the business.

New Market Opportunities

Mostly firms implementing green supply chain management techniques in their supply chain management for grabbing new domestic and international markets. Because green products are acceptable all around the world, most countries do not allow to import polluted products in their countries. Besides, in Pakistan, federal and provincial governments also resist on the polluted products and impose heavy penalties. On the other hand, green products offering firms to attract new markets, which not only improve firms' image in their customers but also increase profits in terms of significant market share.

Reduction in Material Consumption

There is no doubt that green supply chain management is an essential factor in reducing material and chemical costs because the critical objective of green supply chain technique is to reduce harmful effect on the environment and reduce the hazardous chemicals and materials in the production process, which not only enhance

Table 4.28 GSCM motivating factors

S. no.	GSCM drivers/motivating factors
1	Customer pressure
2	Recycle within firm
3	Government rules and legislation
4	Legislation and regulatory compliance
5	Stakeholder pressure
7	Economic benefits of profitable business opportunities
8	Protecting a business from risks of environmental, health, and safety factors
9	Top managerial realization
10	Investor pressure
11	Eco-literacy among supply chain partners
12	Quality products
13	Quality system process
14	Lower rejection and defective products
15	Management commitment towards quality process improvement
17	Warranties and claim policies
18	Price-performance value
19	Compliance with sectorial price behavior transportation cost and production cost
20	Supplying capability
21	The capability of product development order fulfill rate and lead time
22	The capability of R&D technology level
23	High/fast service response
24	Rate of processing order from rate of delivery in time
25	Environment protection system certification
26	Willingness towards investment
27	Benchmarking system
28	Performance measurement system
29	Organizational structure
30	GSCM techniques and methods
31	Interdepartment cooperation
32	Cross-functional team
33	Green design
34	Green manufacturing processes
35	Employee empowerment and motivation towards GSCM
36	Environmental education and training
37	Green organizational culture
38	Trustworthy teamwork
39	Employee involvement
40	Skilled professionals
41	Ethical standards and corporate social responsibilities
42	Supplier commitment
43	Customer feedback

(continued)

Table 4.28 (continued)

S. no.	GSCM drivers/motivating factors
44	Effective communication platform within companies and with suppliers
45	Cooperation among suppliers
46	Green purchasing
47	Green distribution
48	Environmental policymaking
49	Environmental law enforcement
50	Strict supervision
51	Technical enablers/expertise
52	Information technology infrastructure
53	Integration of system
54	Government commitment and support with encouragement
55	Green image and competitive advantage
56	Competitiveness
57	Availability of energy-saving/usage of renewable energy
58	Employee pressure
59	Pressure from labor unions
60	Shortage of landfill areas
61	Media pressure
62	Continuous monitoring and internal audits
63	Environmental protection plans
64	Research and development relationship with suppliers
65	Pollution control and air emissions
66	Control to wastewater and solid waste
67	Internal reduction of production cost
68	Environmental risk minimization (ISO 50001 and ISO 14001)
69	Pressure from NGOs
70	Social acceptance
71	Government incentives and subsidies
72	Industry associations and competitor's actions
73	Early supplier involvement in green product development
74	ISO 14000 certification
75	Ease in getting finance support/loan from banks/financial institutions
76	The increasing scarcity of superior grades of ores
77	Special tax exemption for ISO-certified firms
78	Greater export opportunities
79	Country's environmental awareness
80	New market opportunities
81	Competitive advantage in adopting green practices
82	Reducing the risk of consumer criticism
83	Collaboration with customers
84	Policies and financial consideration regulatory market forces
85	Sociocultural forces
86	International trade unions

(continued)

Table 4.28 (continued)

S. no.	GSCM drivers/motivating factors
87	Operational performance
88	Environmental performance
89	Reduction in material consumption
90	Reduction in consumption of hazardous materials and chemicals
91	Improvement in firm's performance
92	The decrease in the level of manufacturing costs
93	Increase overall customer loyalty level
94	Long-term profits associated with the adoption of "sustainable" strategies
95	Increased efficiency in production, distribution, and storage
96	Renewable Obligation Certificates (ROCs) and fossil fuel costs
97	Availability of new technology
98	Environmental-related programs and meetings and supplier adoptability
99	Awareness about new technology
100	Cooperation with customers
101	Organizational innovation
102	Financial resources and resource availability
103	Certification of suppliers' environmental management systems
104	Green marketing
105	External integration and supplier appraisal
106	SCM and GSCM skills/ knowledge
107	Supply chain flexibility
108	Customer satisfaction
109	Reusing and recycling materials and packaging
110	Avoid fines/penalties
111	Learning capacity
112	Investment recovery
113	Globalization/internationalization
114	Reusable packages
115	Minimize packaging materials
116	Packages with recyclable contents
117	Materials and energy-conscious products
118	Use life cycle assessment
119	Execution of product-oriented EMS (environmental management system)
120	Government environmental inspections
121	Financial incentives from international organizations
122	Increase market share
123	Product price increase

environmental sustainability but also significantly reduce overall supply chain costs in terms of reduction in material and chemical consumptions. In the context of Pakistan, many firms have adopted green supply chain techniques in their firms for enhancing economic/financial benefits. Furthermore, in the province of Pakistan, KPK (Khyber Pakhtunkhwa) government is proving considerable incentives to the firms due to the reduction of hazardous materials and chemicals in their production plants.

Customer Satisfaction

Earning profit is the final goal for all commercial firms, and earning profits is significantly associated with repeat buying, which has a strong relationship with customer satisfaction. In the context of Pakistan, many firms are adopting green practices for the sake of customer satisfaction because customer satisfaction is the final key to success or handsome profit.

Ethical Reason to Increase the Product Price

This is an essential and influential motivating factor in implementing green supply chain techniques in firms. Because after adopting green practices and/or manufactured green products, firms have an ethical reason to increase their prices, which bring improvement in firms' financial performance. On the other hand, regulatory authorities and consumers also have no such objection because firms easily defend their price increment in terms of expensive green materials, and it is a solid reason to increase their products' prices.

4.6.2.2 Organizational Performance Motivating Factors

The organizational performance motivating factors cover the willingness to enhance risk management due to potential disruptions in the SC. The organizational performance drivers mainly cover government rules and legislation, top managerial realization, organizational structure, willingness towards investment, management commitment towards quality process improvement, green organizational culture, organizational-level GSCM techniques and methods, and organizational innovations. In Table 4.30, we can see the complete list of organizational performance drivers, while we will discuss some of them in detail for the development and clarity of understanding.

Table 4.29 Economic performance drivers

S. no.	Economic performance drivers
1	Customer pressure
2	Recycle within firm
3	Stakeholder pressure
4	Economic benefits of profitable business opportunities
5	Investor pressure
6	Lower rejection and defective products
7	Compliance with sectorial price behavior transportation cost and production cost
8	Availability of energy-saving/usage of renewable energy
9	Shortage of landfill areas
10	Internal reduction of production costs
11	Government incentives and subsidies
12	Early supplier involvement in green product development
13	Ease in getting finance support/loan from banks/financial institutions
14	Special tax exemption for ISO-certified firms
15	Greater export opportunities
16	New market opportunities
17	International trade unions
18	Operational performance
19	Reduction in material consumption
20	Reduction in consumption of hazardous materials and chemicals
21	The decrease in the level of manufacturing costs
22	Long-term profits associated with the adoption of "sustainable" strategies
23	Increased efficiency in production, distribution, and storage
24	Renewable Obligation cCertificates (ROCs) and fossil fuel costs
25	Financial resources and resource availability
26	Supply chain flexibility
27	Customer satisfaction
28	Reusing and recycling materials and packaging
29	Avoid fines/penalties
30	Investment recovery
31	Globalization/internationalization
32	Reusable packages
33	Minimize packaging materials
34	Packages with recyclable contents
35	Lower consumption of materials and energy products
36	Financial incentives from international organizations
37	Increase market share
38	Ethical reason to increase the product price

Legislation and Regulatory Compliance

Legislation and regulatory compliance are a significant motivating factor in the adoption of green supply chain practices. Firms adopt green practices due to governmental and regulatory legislation. From the perspective of Pakistan, during the last couple of years, federal and provincial governments have stringent environmental laws, and if any firm violates laws, they face heavy penalties and also lose their reputation.

Top Management Realization

Top management to enforce green practices and techniques in supply chain management plays a vital role. Undeniably, top management has the authority to dictate their ideology over organization policies and practices. If senior management wants to adopt green practices in the supply chain, it is easy to adapt and change all the traditional/polluted processes towards green processes.

Organizational Structure

Organizational structure plays a role as cornerstone, firms processes if easy moldable to green. So it becomes easy for management to adopt green practices and implement new green technology, which can enhance environmental sustainability but also improve organizational performance. Most times, firms would like to adopt green supply chain practices in their operations, but because of the rigid polluted process, it becomes challenging for consultants/specialists in converting polluted systems into the green system. Besides, due to the rigid organizational structure, firms need to restructure all the processes once again for the implementation of green practices, which is a big challenge and needs massive investment. In the context of Pakistan, several firms have adopted green practices in their supply chain management due to the flexible organizational structure.

Employee Involvement

For the adoption of green supply chain methodology, employee involvement plays a role as a motivating factor. Firms cannot implement any plan without the moral acceptance of employees. In simple words, employees are on the fieldwork, and they are an implementer of green practices in the end-to-end supply chain. If employees have a lack of knowledge and awareness regarding green supply chain management, then it is very challenging for firms to implement the green project, and many times results become a failure.

Table 4.30 Organizational performance drivers

S. no.	Organizational performance drivers
1	Government rules and legislation
2	Legislation and regulatory compliance
3	Top managerial realization
4	Lower rejection and defective products
5	Management commitment towards quality process improvement
6	Price-performance value
7	The capability of product development order fulfill rate and lead time
8	Willingness towards investment
9	Benchmarking system
10	Performance measurement system
11	Organizational structure
12	Organizational-level GSCM techniques and methods
13	Green organizational culture
14	Trustworthy teamwork
15	Employee involvement
16	Customer feedback
17	Effective communication platform within companies and with suppliers
18	Strict supervision
19	Employee pressure
20	Pressure from labor unions
21	Media pressure
22	Continuous monitoring and internal audits
23	Environmental protection plans
24	Research and development relationship with suppliers
25	Pressure from NGOs
26	Industry associations and competitor's actions
27	The increasing scarcity of superior grades of ores
28	Policies and financial consideration regulatory market forces
29	Improvement in firm's performance
30	Awareness about new technology
31	Organizational innovation
32	External integration and supplier appraisal
33	SCM and GSCM skills/knowledge
34	Customer satisfaction
35	Learning capacity
36	Globalization/internationalization
37	Reusable packages
38	Use life cycle assessment

Pressure from Labor Unions

Business unions and/or labor unions create pressure on firms to implement green practices in their manufacturing processes for the sake of society and the environment. Most times, firms do not want to adopt green practices due to the enormous fixed costs involved. However, firms also cannot resist the union's pressure. In the context of Pakistan, business and labor unions are very strong more than senior management; if firms ignore their demand, they may stop all the production/manufacturing of firms. Because all labors of firms are a member of the labor union, ultimately, firms have no option except to follow the instructions of labor and business unions.

Environmental Protection Plans

Undeniably, environmental protection plans do not let enterprises violate environmental laws and social concerns. Most federal projects require the development of an environmental protection plan (EPP). An EPP is designed to reduce the impact of the construction and production process on the environment. There is no doubt that small firms have no such environmental protection plan. Nevertheless, multinational firms and corporate sectors have an environmental protection plan, and they review their plan on a yearly and/or quarterly basis. Environmental protection plans ethically motivate firms to adopt more and more green practices in their operations for environmental and social concerns.

Effective Communication Platform Within Companies and with Suppliers

Effective communication between manufacturer and suppliers is essential for the adoption of green projects because most of the times, a supplier may have worked with other green manufacturers and they can provide valuable suggestions, which will not only reduce the implementation time of a project but also reduce the overall costs. In Pakistan, rarely, the manufacturer creates a relationship with the supplier for green projects due to the "fear of privacy or confidentiality." One outstanding quality of managing suppliers is regular and constant communication. Remember, suppliers are not minded readers. If there is some aspect of business relationship firm feel they could be doing better, you need to let them know. If your goals for the next quarter have suddenly changed, you need to make sure they are aware of that fact as soon as possible. You'll also want to regularly get in touch to help keep your supplier on the same page and to make sure they're keeping up with your needs, especially if you're in the type of industry where those needs are likely to change on a regular basis. One of the cornerstones of any type of relationship is a mutual understanding. The only way you're going to be able to maintain mutual understanding between firm business and supplier at all times is through regular communication practices. On the other hand, the automotive industry has started a collaboration and sharing ideas with their suppliers.

Media Pressure

The pressure of media has a significant influence on the firms' decision-making. Most times, firms have no such plan to change their polluted systems towards green due to the enormous costs and specialized expertise/consultants required. However, media create pressure on firms to take responsibility for environmental and social concerns. In the context of Pakistan, the media lobby is robust and sturdy; in fact, the media may create a positive and negative image of the firms if media started a campaign against the polluting firm, so suppliers and distributors also try to avoid to deal with that manufacturer for their reputation and brand image. On the other hand, firms have a big fear of loose customers, and for private firms, "customers" mean income/profit. So due to the fear of media, firms adopt green practices in the manufacturing, distribution, logistics, and transportation activities to enhance and/or maintain their brand image and reputation.

NGO Pressure

In Pakistan, many NGOs (nongovernment organizations) are working for protecting environmental sustainability, and they also create awareness into consumers through different seminars, workshops, and programs. Firms adopt green supply chain management in firms from the fear of "NGO pressure." These NGOs can destroy the image of firms not only at the domestic level but also the international level. Besides, during 2011, some firms have been closed due to pollution concerns.

4.6.2.3 Environmental Performance Motivating Factors

The environmental performance motivating factors mainly cover societal concern for environmental protection; protecting a business from risks of environmental, health, and safety factors; eco-literacy among supply chain partners; environment protection system certification; environmental education and training; ethical standards and corporate social responsibilities; environmental law enforcement; pollution control and air emissions; control to wastewater and solid waste; environmental risk minimization (ISO 50001 and ISO 14001); and government environmental inspections. In Table 4.31, we can see the complete list of environmental performance drivers, while we will discuss some of them in detail for the development and clarity of understanding.

Societal Concern for Environmental Protection

Many firms are implementing green supply chain management techniques for societal and environmental protection. Undeniably, higher environmental performance indirectly increases economic performance and also builds a competitive advantage for the firms. Besides, customers also buy green products in the sense of firms' sympathy, and government support firms financially and ethically. On the other hand, societal concern for environment protection increases the firm's market share and also allows exporting products in foreign countries with the lowest rejection rate.

Eco- literacy Among Supply Chain Partners

A higher rate of ecological literacy among supply chain partners is the significant driver of green supply chain management. Usually, firms failed in green projects due to limited knowledge of ecological and/or green practices. They only adopt green practices for competing with their competitors. Also, a higher rate of green practice awareness in supply chain partners is essential because the manufacturer buys products from its suppliers and sells to customers. If these two (upstream and downstream) partners have sufficient knowledge and awareness of green practices, material, strategies, it will significantly benefit the manufacturers.

Environmental Law Enforcement

In the context of Pakistan, firms are also adopting green practices in their supply chain management due to environmental law enforcement. There is no doubt that since 2013, federal and provincial government strict their environmental policies and legislation. Besides, if any firm was found to violate the laws, the government will directly impose heavy penalties, including 50000USD or ban the firm's operations in the territory. It is the motivating factor for the firms to adopt green practices in their manufacturing, process, logistics, and distribution activities.

Pollution Control and Air Emissions

Pollution control and air emissions are the motivating factors for the corporate sector, and many enterprises have started a "social corporate responsibility program" to reduce their adverse and/or harmful effects on environmental sustainability. Cost analysis models and tools for air pollution regulations support the assessment of emission reductions and engineering costs for air pollution control strategies. They are addressing stationary sources of air pollution in Pakistan. Learn about reducing emissions of hazardous air pollutants in Pakistan.

Environmental Risk Minimization (ISO 50001 and ISO 14001)

ISO 50001 and ISO 14001 are the motivating factors for implementing green supply chain practices in firms' operations. In the context of Pakistan, several firms are adopting green practices, and the key motivating factors are these two ISOs.

Government Environmental Inspections

Since 2013, federal and provincial governments conduct a surprise inspection of firms' processes and plant-surrounded areas to find whether firms are working on environmental sustainability and fulfilling the environmental laws. Government inspections increase the rate of implementing green projects in Pakistan; besides if firms took all the essential measures to secure environmental sustainability, so the government is also offering incentives and appreciations, which have a significant impact on firms marketing strategies.

Control to Wastewater and Solid Waste

Reduction in wastewater and solid waste also means a reduction in production and purchasing costs, which majorly motivates firms to adopt green practices in their supply chain management. Many firms have been noted, which adopt green practices only to reduce their waste, which is, directly and indirectly, correlated with firms' profitability and increase of overall system efficiency.

Protecting a Business from Risks of Environmental, Health, and Safety Factors

Firms are implementing green supply chain methodologies in their operations to reduce environmental risks such as BP (British Petroleum) has suffered in the last decade due to no environmental severe risk measures. British Petroleum had a vast network in Pakistan for long decades, but once their oil container leakage in the sea and all the oil destroy to the environmental sustainability in terms of heavy oil floated on the sea, and thousands of sea animals died. Finally, the government imposes heavy penalties in thousands of billions of dollars, and British Petroleum became out of business. "One mistake can be the last mistake," so, in Pakistan, firms are seriously adopting green practices to reduce environmental, safety, and health risks.

Environmental Education and Training

There is no doubt that education and training change people's lives and thinking way. As we have discussed, since 2013, the government is seriously working on environmental sustainability, such as the "Billion Tree Tsunami Project" conducted by the provincial government. Besides, federal and provincial governments also conduct seminars and workshops in different public forums, including university level to increase environmental-friendly policy needs and knowledge. On the other hand, several institutions also have announced a unique subject, "Green Supply Chain Management," for graduate and undergraduate students.

Table 4.31 Environmental performance drivers

S. no.	Environmental performance drivers
1	Societal concern for environmental protection
2	Protecting a business from risks of environmental, health, and safety factors
3	Eco-literacy among supply chain partners
4	Environment protection system certification
5	Environmental education and training
6	Ethical standards and corporate social responsibilities
7	Green purchasing
8	Environmental policymaking
9	Green distribution
10	Environmental law enforcement
11	Integration of systems
12	Government commitment and support with encouragement
13	Pollution control and air emissions
14	Control to wastewater and solid waste
15	Environmental risk minimization (ISO 50001 and ISO 14001)
16	ISO 14000 certifications
17	Country's environmental awareness
18	Reduction in consumption of hazardous materials and chemicals
19	Environmental-related programs and meetings and supplier adoptability
20	Certification of suppliers' environmental management systems
21	Minimize packaging materials
22	Packages with recyclable contents
23	Lower consumption of materials and energy products
24	Execution of product-oriented EMS (environmental management systems)
25	Government environmental inspections

Ethical Standards and Corporate Social Responsibilities

Many firms are already working on corporate social responsibilities, but it does not mean that they have complete green supply chain operations in their firms. There is no doubt that for the last couple of years in Pakistan, which firms already were working on corporate social responsibility, they quickly adopted green supply chain techniques and methods due to their more excellent knowledge in environmental sustainability, customer mindset, and government legislation of environment. Corporations can have enormously detrimental effects on the environment. Companies can invest in local communities in order to offset the negative impact their operations might have. A natural resource firm that begins to operate in a poor community might build a school, offer medical services, or improve irrigation and sanitation equipment.

Similarly, a company might invest in research and development in sustainable technologies, even though the project might not immediately lead to increased profitability. Oil spills are some of the most conspicuous examples, but industries, as varied as chemical manufacturing, mining, agriculture, and fishing, can do permanent damage to local ecosystems. Climate change can also be attributed in large part to corporations. While their responsibility is hard to untangle from that of the consumers who demand electricity and transportation, it is difficult to deny that many corporations have profited from the deterioration of the global environment.

Competitive Advantage Motivating Factors

The competitive advantage motivating factors mainly covers quality systems process capability and quality assurance, fast service response, rate of processing orders from the rate of delivery in time, skilled professionals, technical enablers/expertise, research and development relationship with suppliers, green manufacturing processes, reducing risk of consumer criticism, and green marketing. In Table 4.32, we can see the complete list of competitive advantage drivers, while we will discuss some of them in detail for the development and clarity of understanding.

Quality Systems Process

Most firms adopt green supply chain practices for enhancing their quality of products and systems capability to grasp competitive advantage in the market. There is no doubt that green supply chain practices enhance the quality of products and processes due to the elimination of waste in the systems. Quality assurance (QA) is a way of preventing mistakes or defects in manufactured products and avoiding problems when delivering solutions or services to customers, which ISO 9000 defines as "part of quality management focused on providing confidence that quality requirements will be fulfilled." From the perspective of Pakistan, several enterprises successfully implemented green supply chain practices in their processes and logistics operations to gain a competitive advantage.

Fast Service Response

Green supply chain practices are somehow the same as lean manufacturing, and they also eliminate the waste from the systems, which considerably enhances the speed of service. Untimely response to customer requests is one of the hallmarks of poor customer service. Customers want to feel valued; they want to know their business is appreciated. Therefore, customers are likely to take their business elsewhere

Table 4.32 Competitive advantage drivers

S. no.	Competitive advantage drivers
1	Quality products
2	The capability of R&D technology level
3	Supplying capability
4	Quality systems process
5	Lower rejection and defective products
6	Fast service response
7	Rate of processing orders from the rate of delivery in time
8	Interdepartment cooperation
9	Green design
10	Green manufacturing processes
11	Cross-functional team
12	Employee empowerment and motivation towards GSCM
13	Skilled professionals
14	Supplier commitment
15	Cooperation among suppliers
16	Green purchasing
17	Green distribution
18	Technical enablers/expertise
19	Information technology infrastructure
20	Green image and reputation
21	Research and development relationship with suppliers
22	Social acceptance
23	Early supplier involvement in green product development
24	Competitive advantage in adopting green practices
25	Competitiveness
26	Reducing the risk of consumer criticism
27	Collaboration with customers
28	Sociocultural forces
29	Increase overall customer loyalty level
30	Availability of new technology
31	Cooperation with customers
32	Green marketing
33	External integration and supplier appraisal
34	SCM and GSCM skills/knowledge
35	Supply chain flexibility
36	Customer satisfaction
37	Reusing and recycling materials and packaging
38	Avoid fines/penalties
39	Learning capacity
40	Reusable packages
41	Investment recovery
42	Increase market share
43	Ethical reason to increase the product price

if they feel neglected, if they have an immediate need for a service that you are failing to deliver, or if they do not appreciate your work ethics. Poor response time, especially if done repeatedly, results in loss of customers and revenue. The key to generating loyal customers is to provide them with efficient service by the required time frame. For example, if you provide in-house electronic repair services, arrive on time for appointments, and keep the lines of communication open between you and the customer. Besides, the green supply chain's main objective is to reduce hazardous and toxic materials from the system to enhance environmental sustainability, but indirectly it also improves firms' financial and customer satisfaction KPI (key performance indicators).

Skilled Professionals

Firms cannot adopt green supply chain management practices without skilled professionals' employees and consultants. Green practices and projects require an experienced team to understand the product, process, and organizational structure and accordingly suggest how to implement green practices.

Green Image and Reputation

Green image and reputation are some of the significant reasons to adopt green supply chain practices in firms' operations. Most companies adopt green logistics, green distribution, green transportation, green manufacturing, and green purchase for enhancing their positive image and reputation in the market to increase market share and profitability.

Green Marketing

Firms adopt green practices to enhance their green marketing techniques to increase profitability and market share. In modern times, green marketing is a desirable and successful strategy to increase profitability with environmental performance. In the context of Pakistan, firms adopt green practices to create a competitive advantage over their competitors and earn healthy profitability.

References

Al-Hindawe, J. (1996). Considerations when constructing a semantic differential scale. *La Trobe Working Papers in Linguistics., 9*, 41–58.

Anderson, J., & Gerbing, D. W. (1988). Structural equation modeling in practice: A review and recommended two-step approach. *Psychological Bulletin, 103*(3), 411–423. https://doi.org/10.1037/0033-2909.103.3.411

Armstrong, J. S., & Overton, T. S. (1977). Estimating nonresponse bias in mail surveys. *Journal of Marketing Research, 14*, 396–402. https://doi.org/10.2307/3150783

Bentler, P. M., & Bonett, D. G. (1980). Significance tests and goodness of fit in the analysis of covariance structures. *Psychological Bulletin, 88*(3), 588–606. https://doi.org/10.1037/0033-2909.88.3.588

Chin, W. W. (1998). The partial least squares approach for structural equation modeling. In G. A. Marcoulides (Ed.), *Methodology for business and management. Modern methods for business research* (pp. 295–336). New Jersey: Lawrence Erlbaum Associates Publishers.

Chin, W. W. (2010). How to write up and report PLS analyses. In V. Esposito Vinzi, W. W. Chin, J. Henseler, & H. Wang (Eds.), *Handbook of partial least squares: Concepts, methods, and applications* (pp. 655–690). Heidelberg, Heidelberg/Dordrecht, the Netherlands/London/New York: Springer. https://doi.org/10.1007/978-3-540-32827-8_29

Chin, W. W., & Newsted, P. R. (1999). *Structural equation modeling analysis with small samples using partial least squares, statistical strategies for small sample research.* London: Sage Publications.

Cohen, J. (1992). A power primer. *Psychological Bulletin, 112*(1), 155–159. https://doi.org/10.1037//0033-2909.112.1.155

Diamantopoulos, A., & Windlhofer, H. (2001). Index construction with formative indicators: An alternative to scaling development. *Journal of Marketing Research, 38*(2), 269–277.

Dijkstra, T. K., & Henseler, J. (2015). Consistent partial least squares path modeling. *MIS Quarterly, 39*(2), 297–316.

Fornell, C., & Larcker, D. F. (1981). Evaluating structural equation models with unobservable variables and measurement error. *Journal of Marketing Research, 18*(1), 39–50. https://doi.org/10.2307/3151312

Garver, M. S., & Mentzer, J. T. (1999). Logistics research methods: Employing structural equation modeling to test for construct validity. *Journal of Business Logistics, 20*(1), 33–57.

Gefen, D., Straub, D., & Boudreau, M. (2000). Structural equation modeling and regression: Guidelines for research practice. *Communications of the Association for Information Systems, 4*, 1–77. https://doi.org/10.17705/1CAIS.00407

Hair, J., Anderson, R., Tatham, R., & Black, W. (1998). *Multivariate data analysis* (5th ed.). New Jersey: Prentice-Hall.

Hair, J. F., Christian, M. R., & Marko, S. (2011). PLS-sem: Indeed, a silver bullet. *The Journal of Marketing Theory and Practice, 19*(2), 139–151.

Hair Jr., J. F., Black, W. C., Babin, B. J., Anderson, R. E., & Tatham, R. L. (2006). *Multivariant data analysis*. Upper Saddle, NJ: Pearson International Edition.

Henseler, J., Ringle, C., & Sinkovics, R. (2009). The use of partial least squares path modeling in international marketing. In R. Sinkovics & P. Ghauri (Eds.), *New challenges to international marketing (Advances in international marketing, Vol. 20)* (pp. 277–319). Bingley, England: Emerald Group Publishing Limited. https://doi.org/10.1108/S1474-7979(2009)0000020014

Henseler, J., & Sarstedt, M. (2013). Goodness-of-fit indices for partial least squares path modeling. *Computational Statistics, 28*(2), 565–580.

Hoelter, J. W. (1983). The analysis of covariance structures: Goodness-of-fit indices. *Sociological Methods and Research, 11*, 325–344.

Hu, L.-T., & Bentler, P. M. (1999). Cutoff criteria for fit indexes in covariance structure analysis: Conventional criteria versus new alternatives. *Structural Equation Modeling, 6*(1), 1–55. https://doi.org/10.1080/10705519909540118

Kline, R. B. (2005). *Methodology in the social sciences. Principles and practice of structural equation modeling* (2nd ed.). New York: Guilford Press.

Lohmoller, J. B. (1989). *Latent variable path modeling with partial least squares*. Heidelberg, Heidelberg: Physica-Verlag. https://doi.org/10.1007/978-3-642-52512-4

Lowry, P. B., & Gaskin, J. (2014). Partial least squares (PLS) structural equation modeling (SEM) for building and testing behavioral causal theory: When to choose it and how to use it. *IEEE Transactions on Professional Communication, 57*(1–2), 123–146. https://doi.org/10.1109/TPC.2014.2312452

Ma, M., & Agarwal, R. (2007). Through a glass darkly: Information technology design, identity verification, knowledge contribution in online communities. *Information Systems Research, 18*(1), 25–31.

Nunnally, J. C. (1978). *Psychometric theory* (2nd ed.). New York.: McGraw-Hill.

Podsakoff, P. M., Mackenzie, S. B., Lee, J. Y., & Podsakoff, N. P. (2013). Common method biases in behavioral research: A critical review of the literature and recommended remedies. *Journal of Applied Psychology, 88*(5), 879–903.

Podsakoff, P. M., & Organ, D. W. (1986). Self-reports in organizational research: Problems and prospects. *Journal of Management, 12*(4), 531–544. https://doi.org/10.1177/014920638601200408

Ringle, C. M., Wende, S., Will, A. (2005). SmartPLS 2.0. http://www.smartpls.de

Schumacker, R. E., & Lomax, R. G. (2004). *A beginner's guide to structural equation modeling* (2nd ed.). New Jersey: Lawrence Erlbaum Associates Publishers.

Tabachnick, B. G., & Fidell, L. S. (2001). *Using multivariate statistics* (4th ed.). Needham Heights, MA: Allyn & Bacon. ISBN 0-321-05677-9. Hardcover.

Tenenhaus, M., Vinzi, V. E., Chatelin, Y. M., & Lauro, C. (2005). PLS path modeling. *Computational Statistics & Data Analysis, 48*, 159–205. https://doi.org/10.1016/j.csda.2004.03.005

Wetzels, M., Odekerken-Schroder, G., & Van Oppen, C. (2009). Using PLS path modeling for assessing hierarchical construct models: Guidelines and empirical illustration. *MIS Quarterly, 33*, 177–195.

Chapter 5
Discussion on Green Supply Chain Management

5.1 Quantitative Part of the Research

First of all, we will discuss the quantitative part of this research through the estimation results of modeling. As the model developed to prove that green supply chain management practices enhance competitive advantage, economic performance, environmental performance, and organizational performance. Undeniably, this research proves that these drivers are playing a significant role in the adoption of green supply chain management practices. The following discussion has been mentioned based on hypothesis results.

5.1.1 IEM and GIS

During the testing of the hypothesis, the research found that internal environmental management has a positive and significant effect on green information systems. It means that due to the adoption of internal environmental management, firms have flexibility and adaptability to implement end-to-end green supply chain management, which will further affect the firms' performance. The contribution of internal environmental management into the implementation of green information systems is very significant. As per the statistical tests, a 1% increase in the internal environmental management will improve the green information systems by 0.73%. Gupta and Barua (2017), the success of green information programs in any firm are depending on the top management mentality. Because senior management decides whether firms should adopt green practices to protect environmental sustainability or non-green practices from earning only money.

Environmental sustainability issues in the sector of FMCG have attracted worldwide concerns in the last couple of years. Firms and regulatory authorities have

S. A. R. Khan, *The Critical Success Factors of Green Supply Chain Management in Emerging Economies*, EAI/Springer Innovations in Communication and Computing, https://doi.org/10.1007/978-3-030-42742-9_5

started to realize the significance of greening their end-to-end supply chain and logistics functions. The green supply chain has achieved special attention of firms, and companies are very interested in implementing green strategies and practices into their business operations. The idea of this research is to examine the significant performance drivers for green supply chain execution. After in-depth literature surveys and consulting with academic and industry experts, the research has advised 13 essential performance indicators. The phase of quantitative was scheduled through a broad survey using a structured questionnaire with many agro-based firms followed by a phase of qualitative, and the research has concluded that IEM regulatory pressure and environmental design are well-linked together, and they are significantly affecting on the firms' performance.

Research results show that the adoption of the green supply chain will not only enhance the performance of the environment but also enhanced and improved the financial performance of the firms. Further, research suggests that firms need to adopt internal environmental management systems to enforced green information systems into their supply chain operation, which may improve overall firms' performance, including operational, environmental, and financial. Besides, internal environmental management creates pressure on firms to adopt green purchasing, green logistics, green distribution, and green packaging, which significantly support the green supply chain phenomena and enhanced firms' competitive edge in the market. Scur and Barbosa (2017) researched on home appliance industry of Brazil; the idea of this research is to examine and analyze the ground practices of GSCM implemented by home appliance firms. Researchers used the following variables to estimate the effect of internal environmental management: top management commitment to GSCM, internal environmental audit programs, ISO 14001 certification, and staff cooperation for environmental enhancements.

Environmental concerns in the industry of mining require holistic and systematic approaches with internal environmental management to support these problems. IEM systematic adoption offers auditing and monitoring for environmental compliance of mining operations. Environmental concerns in the mining industry require systematic and holistic approaches with suggestions, senior management, and incorporation of TQM into internal environmental systems which may support to reduce suppliers' environmental degradation. In the internal environmental management systems, firms should mainly examine total quality environment management, pollution prevention plans, ISO 14001 certification by suppliers, and employee incentive programs (Kusi-Sarpong, Sarkis, & Wang, 2016). Seong-Tae Kim Sang-Yoon Lee (2012) researched to find the relationship between internal environmental management, stakeholder pressure, and organizational performance from the perspective of South Korea. The results show that internal environmental management has a positive and significant effect on the adoption of green practices in the supply chain and logistics, while stakeholder pressure may create the positive effect but sometime may also affect negatively on the adoption of green practices due to unnecessary pressure on the manufacturing firms or financial scarcity/constraint.

The definition of internal environmental management is "the practice of developing sustainability of the environment as a strategic firm imperative through support and commitment of the imperative from mid-level managers and senior-level managers." A GIS (green information systems) shows the backbone for efforts of environmental management by supporting the company's internal system of management by fulfilling the needs of many stakeholders mainly including customer pressure, suppliers, governmental regulatory, and investors. The system of internal environmental management does not directly affect environmental collaboration with customers. IEM significantly effect on GIS (green information systems) and environmental collaboration with suppliers, while indirectly affect environmental collaboration with customers through green information systems.

Internal environmental management and GIS are recognized as critical precursors to the adoption of cooperation with customers, green purchasing, investment recovery, and eco-design. Once the company implements environmental sustainability as a strategic imperative, then firms may proceed to grow GIS capabilities. Information systems are critical for the maintenance, creation, and survival of SCs. Further, IS not only allows systems to be interconnected; but they also can be used to improve commitment, collaboration, and trust among supply chain stakeholders. IS is primarily used by firms for offering techniques, tools, and mechanisms for trust development.

IEM provides the firms support to green supply chain management endeavors through managerial control, policymaking, and environmental auditing. Internal environment management leads to top management, zero-tolerance, and interdepartment collaboration and coordination. Further, internal environmental management significantly contributes to the enhancing of green information systems in the supply chain environment. Giovanni and Esposito (2012) researched to examine the association between environmental management and firm performance to verify whether the adoption of an effective IEM is a company's initial condition to relate with green supply chain and which type of environmental practices significantly participate the most in enhancing organizational performance. Researchers used SEM (structural equation modeling) method for testing hypotheses on Italian companies and calculated the structural paths between constructs by component-based approaches and means of both variances. After the usage of estimation methods enable to contributing to the current debate about the specification of the firm performance. The results of the research show that being internal environmental management is an initial condition to adopt green supply chain practices; further internal environmental management creates pressure to implement green information systems, which can positively affect the environmental performance and financial performance along with the operational performance of the firms.

5.1.2 *IEM and GM*

The research found that internal environmental management has a positive effect on green manufacturing. In simple words, international environmental management encourages firms to adopt green manufacturing operations and green practices in their supply chain management. The effect of international environmental is significantly positive on green manufacturing, while results show that a 1% increase in the internal environmental management activities will improve green manufacturing practices and operations by 0.39%, which will also create an effect on the firms' overall performance. GSCM has emerged as an effective and efficient management philosophy for leading manufacturing firms. Undeniably, the People's Republic of Pakistan is a developing country with greater manufacturing operations, but still greener products market is narrow as compared to developed countries.

In the last couple of years, Pakistani manufacturers have started thinking about green approaches and practices to implement in their supply chain and logistics activities, but the key attraction for Pakistani producers is building competitive advantage with enhanced environmental and financial performance. Sustainable supply chain management has attracted increased attention in the last couple of decades. The surveys were arranged based on feedback from the industries in Pakistan.

Researchers found that green supply chain management practices in Pakistan are still in its growing phase, the awareness of green products is better than before, and customers started to buy greener products in facts if its little expenses too. The findings show that collaboration for environmental sustainability has a significant and positive effect on a healthy environment, which also creates a competitive advantage for the firms and companies to increase their overall profitability in terms of waste reduction, elimination, value stream mapping, and total quality management. Furthermore, the green supply chain market is growing speedily all around the globe. Many countries started to green their business operations, and customers are also well aware of environmental sustainability. Due to the higher pressure of government and customers, firms are shifting their polluted supply chain over the green and clean supply chain. On the other hand, many developing and developed countries, including the USA, Germany, France, the UK, and Pakistan, already adopted green practices in their supply chain activities to protect environmental sustainability and reduce harmful effects on human lives.

The laws of air quality are usually explicitly designed to protect people's health by eliminating airborne. Other initiatives are designed to address broader green issues like a limitation of carbon emissions to protect the ozone layer, climate change, and acid rain. The efforts of regulatory authority are identifying and limiting air pollutants and enforcing necessary and suitable mitigation technologies. The quality of water laws protected sea pollution, including ground, surface, and stored drinking water. Many qualities of water laws, such as drinking water regulations, can be designed concerning people's health. On the other hand, chemicals and a colossal waste go into the sea and river, which create destructive situations for human lives but also have severe consequences for flora and fauna lives.

One key reason for environmental degradation is industrialization; the government alone cannot protect environmental sustainability; and public awareness plays a significant role in controlling pollution and climate change issues. Environmental education is critical to encourage customers and firms to take environmental friendly initiatives in their society. It sensitizes the society regarding environmental challenges and issues of interested individuals to build expertise and skills, thereby offering suitable solutions. Loss of biodiversity, ozone layer depletion, climate change, declining fisheries, illegal trade of endangered species, land degradation, destruction of habitats, depletion of groundwater supplies, environmental pollution, stormwater, and sewage disposal pose a severe threat to green systems in urban, forest, rural, and marine ecosystems. Undeniably, informal and formal education about the environment will give interested individuals the knowledge, skills, values, and tools needed to solve the environmental challenges on a domestic and international level.

The laws of the environment are also called natural resources law; the term describes the network of statutes, treaties, regulations, and customary laws addressing the impact of human activities on environmental sustainability. The critical law of environmental regimes addresses environmental pollution. An associated but different set of regulatory regimes, now strictly influenced by environmental legal philosophies, emphasize on the management of specific resources, including minerals, forests, fisheries, etc. The quality of air has been destroyed due to heavy pollution in the environment. Zhu, Sarkis, and Lai (2008) researched to investigate the association between green practices and their impact on firms' performance. Researchers used some environmental factors, including green purchasing, cooperation with customers, eco-design, and investment recovery, while the findings show that green practices are not as such, improving the financial performance of the firms.

On the other hand, green practices and approaches are significantly enhancing environmental performance and creating a competitive edge in the international market. Liu, Tang, and Xue (2012), based on a survey of production companies in Central and South China, investigate the association between GSCM, firm performance, and competitive edge. By using factor analysis techniques, the scale of manufacturing companies' GSCM is obtained. The finding shows that GSCM can be conceptualized as four key elements: green purchasing/procurement, green logistics, green manufacturing, and green distribution. Further, the researcher used LISREL software to test the hypothesis. The results show that green distribution, green manufacturing, and green logistics have a positive and significant impact on company performance.

Generally, the laws of waste designed to reduce the uncontrolled dispersal of waste materials into the atmosphere in a way that may cause environmental problems and pollution also encourage recycling and re-manufacturing and reduce approaches of green supply chain management and green practices in business. The efforts of the government include identifying and classifying the types of waste and their treatment and disposal practices. Jakowczyk, Quariguasi, Neto, Gibson, and Van (2017) researched understanding the market for re-manufactured products. This research paper is to cover the gap regarding re-manufactured products in the France market. Emphasizing on the online market for re-manufactured electronic items, researchers empirically test whether market-specific and product-specific determinants impact intensity. Researchers collected the data from online sources, and the results indicate that the number of re-manufactured items listed is correlated with several listing for two product-specific characteristics. The research focused on senior management commitment towards green practices in the supply chain operations, and the findings show that green practices are significantly increasing overall firms' performance. On the other hand, green practices are also reducing the cost from the systems in terms of stream mapping and total quality management to eliminate waste.

Environmental pollution destroys sustainability and creates significant problems for human lives. The laws of pollution control are designed to solve the issues of the environment and consequently must often explain not only the necessary response actions but also need parties to take initiatives to keep the environment clean and green. The requirement of regulatory authorities may include liability allocation, emergency response, feasibility studies, site assessment, and post-remedial monitoring and remedial investigation. Khan et al. (2017) researched to investigate the causal relationship between environmental logistics indicators and growth. Researchers selected 15 different countries around the world from 2007 to 2015. The researchers used econometric techniques to test the hypothesis. The results show that the per capita income, manufacturing, industry, and service share to gross domestic product are significantly affected by carbon emissions and greenhouse gases. The competence and infrastructure of logistics enhance financial growth in the countries. The casual associations confirm that more demand for energy results in an increase industry value-added, service sector, and economic growth. The manufacturing sector emits a large number of carbon emissions that have a large consumption of resources. The waste manufacturing rate far exceeds the global recovery capability and compensation, causing the global ecological systems to fail to maintain a healthy and sustainable environmental development.

Undeniably, the chemical used in the products severely damages the environmental sustainability and human lives, while flora and fauna are also badly destroyed due to industrialization. The concept of the green supply chain protects not only the environmental sustainability but also the lives of human. The government efforts to protect the environment include banning hazardous chemicals, including polypropylene and black plastic bags, etc. Product innovation and GSCM practices both have a pivotal source to achieve a competitive edge in the market. However, research on the configuration of GSCs for new products shows a comparably new trend. The researcher used a goal programming approach, drawn by the case example of FMCG producers. In this perspective, trade-offs between the ecologic factor of carbon emissions and financial value creation. Kee-hung Lai and Wong (2012) green logistics and their effect on the performance, the value chain of each business firm is increasingly offered to demonstrate the profitability of their basic operations, outbound logistics, inbound logistics, and marketing. The implementation of green logistics shows an opportunity for Chinese producers to export in European countries. The results indicate that logistics green practices, including internal environmental management and green manufacturing, have a strong correlation. Internal environmental management encourages firms to adopt green manufacturing practices, which significantly and positively contribute to the firms' performance.

The human factor is very significantly important for enhancing logistics performance, while infrastructure and institutions are not very important for improving logistics performance. Investments on human are suitable means of the economic outlook, and stimulating innovation perceived that the sector of logistics accounts for an avg. 10% of the whole EU's GDP (gross domestic product). Taylor, Mitra, and Datta (2014) conducted a study on sustainable supply chain management which has attracted increasing attention in the last couple of decades. Although sustainable supply chain management has been in-depth studied for developing countries and developed countries. The surveys were arranged based on feedback from the industries in India. Researchers found that green supply chain management practices in India are still in its beginning; the awareness of green products is not enough, and customers are still focusing on the cheap/low-cost products. The findings show that collaboration for environmental sustainability has a significant and positive effect on the healthy environment, which also create competitive advantage for the firms and companies to increase their overall profitability in terms of waste reduction, elimination, value stream mapping, total quality management, etc. Finally, researcher calls for further research due to unclear picture of green supply chain practices (internal environmental management) on green manufacturing.

5.1.3 IEM and GD

The results indicate that internal environmental management has a positive and significant effect on green distribution. There is no doubt that internal environmental management is the concept and ideology of senior management, which insist on the organization to adopt green distribution practices in its supply chain management. The finding shows that a 1% increase in internal environmental management will improve green distribution practices by 0.27%. Besides, senior management has a very positive effect on the organizational strategy and supply chain strategy to mold them towards the green side. A wide variety of issues regarding environmental-friendly transportation, such as the enhancing of alternative options such as renewable energy, fuels, green intelligent transportation systems, next-generation electric vehicles.

The route of vehicles and better utilization of cars would more effectively lead to green transportation schemes. For the last couple of decades, the critical challenge is to design a network of green distribution using vehicle routing models. The adoption of GSCM may be evaluated by assessing a set of specific elements to capture the difficulty observed in representative practices. The current research is to evaluate a model for evaluation of the helpful adoption of green supply chain management in industrial phenomena of supply chain management: drivers, barriers, green distribution, and ecological design. Improvement effects towards green supply chain management, conducted by top management, should emphasize on the practices of constructs with green innovation and strategy. The impact of environmental assessment is the assessment of consequences which means a positive and/or negative impact on the environment. EIA (environmental impact assessment) is typically used when applied to real projects by firms and/or individuals and the term "strategic environmental assessment" applied to plans, policies, and programs.

The resources of water and mineral are essential factors for country development and growth. The mining is affected by many regulations regarding the safety and health of miners, as well as the environmental impact of carbon emissions, GHG (greenhouse gases), and global warming. Liu et al. (2012) conducted the research impact of green practices on firms' performance and their competitive advantage. The distribution with green practices comprises the efficiency of the distribution system about environmental management and environmental-friendly consumerism: the environmental-friendly product recycling and their packing and packaging. The results show three key conclusions; first, the scale of GSCM is developed; green supply chain management of manufacturing companies in China can be conceptualized as a four-dimensional construct including green purchasing, green logistics green manufacturing, and green distribution. Second, the four elements of green supply chain management are positively correlated with firms' competitive edge. Third, green logistics and green manufacturing increased the firms' performance.

The resources of the forest protection environmental sustainability keep the air quality good for human lives. However, unfortunately, the forest is declining due to industrialization, and carbon emissions are also increased in a dangerous position. Forest absorbed air pollution (carbon dioxide) and emitted oxygen, which protects the human, flora, and fauna lives. Mainly it is the government and regulatory authorities work to protect trees and to plant trees, but due to greater public awareness, different kinds of NGOs and individual people are also planting trees. Lau (2011) researched the green performance of logistics, including green material management, green purchasing, green marketing, and green distribution. The key objective is to minimize the effect of logistics operations and activities on environmental sustainability, improve the value of the product, and lower the cost of manufacturing. The findings prove that green logistics operations may achieve minimum inventory level with minimized logistics cost and enhanced customer service with improved firms' image and reputation in the market. Kim and Min (2011) logistics operations may have a significant effect on carbon emissions and greenhouse gases; it was found that many countries select to improve per capita income/GDP at the expense of environmental degradation. Accordingly, the green logistics performance index is suggested as a good indicator of the green efficiency of logistics, indicating that affects the logistics competitiveness of a country has on its environment.

The process of recycling is to convert waste materials into useable form and shape. The recycling process and re-manufacturing process are important steps in the concept of GSCM "green supply chain management." Waste disposal can reduce the cost of products and also increase environmental sustainability. Furthermore, the recycling idea is vastly used in many countries, including Pakistan, the USA, the UK, and Germany, to control climate change, air pollution, and water pollution. The process of recycling has some ISO standards such as ISO 15270:2008 for the waste of plastic and ISO 14001:2004 for environmental management control. Recycling may include many kinds of items, including cardboard, tires, plastic, paper, glass, and electronic items, and each item has their recycling process. The primary procedure of recycling includes collecting, sorting, cleaning, and re-manufacturing. The primary two types of recycling are one to the recycling of material and produce a new product; for example, used plastic bottles and converted into different shapes of plastic bottles. However, this is often difficult or too expensive (compared with manufacturing the same item from raw materials/components), so mainly recycling of several products involves their reuse in manufacturing different materials or due to their hazardous nature such as thermometers, etc. (Ben Moussa, Rasovska, Dubois, De Guio, & Benmoussa, 2016) green SCM contain a wide range of green practices classified into three major categories: green product, green operations, and managerial practices. Green packaging and green product design, which objective is to enhance ecological conditions during the product improvement stage. Green operation's aim is to explore green (Luthra, Govindan, Kannan, Mangla, & Garg, 2017) product, green purchasing, reverse logistics, and green distribution, which split into green transportation and green storage with green distribution network design. Finally, practices of managerial correspond to a set of practices with strategies and methods to manage, coordinate, and design of the supply chain.

(Liu et al., 2012) The study planned on green supply chain practices and their effect on the firms' competitive advantage. The results indicate that green manufacturing logistics and reverse logistics significantly and positively affect the firms' performance and also have a significant role in building a competitive edge. Furthermore, green purchasing and green distribution have no such direct effect on the firms' performance and competitive edge building. Researchers also call for further research to identify the real relationship between green purchasing, green distribution with firms' performance, and competitive advantage. The distribution stage contains all processes, which are compulsory to store and move items on their way from manufacturing to the customers, further consisting of warehousing and transportation activities.

Researchers are dealing with the performance of environmental sustainability with different distribution network designs. Global sourcing accelerates the current boom in the sector of transportation, resulting in greater distances and lead time between origins to destination. In many European Union countries, firms are considerably adopting green practices in their distribution network to reduce the harmful effect on the environment, while regulatory authorities also have stringent laws and regulations; if any firm does not follow environmental policies, they may face legal actions and substantial financial penalties with society boycott. The efficient routing of heavy trucks and cars in a firm's network of distribution results in economic, and many cases, in environmental sustainability advantages. Enhance the distribution systems and transport efficiency processes are made on mid-term planning underlying transport processes (Zhou, Gong, Huang, & Peters, 2017). The researchers indicate that in this work, the significant and positive consequences accurately examine transport processes. Besides, decisions of inventory into routing considerations may also provide better results (Mirhedayatian, Azadi, & Farzipoor Saen, 2014).

5.1.4 IEM and GT

The research results show that internal environmental management has a positive and significant association on a 1% confidence level. Furthermore, results also indicate that a 1% increase in the internal environmental management will lead to a 0.37% increase in transportation cost. Undeniably, senior management commitment has a strong effect on the firms' strategies and their practices to adopt green supply chain operations. On the other hand, the transportation sector has a significant reason for higher carbon emissions and GHG emissions, which not only destroys the environmental sustainability and fauna and flora but also increased to the overall supply chain cost of the firms. The polluted transportation modes are the "evil," which have nothing good, but only bad, and polluted transportation increased overall firms' cost, reduces environmental sustainability, and creates global warming issues.

The environmental situation in Asian countries is also not very good, and due to these reasons, many Asian countries built a union called AECEN (Asian environmental compliance and enforcement network). In this union, a total of 16 Asian countries are actively playing a role such as Pakistan, China, Cambodia, Indonesia, Maldives, India, Japan, Malaysia, Korea, the Philippines, Nepal, Sri Lanka, Singapore, Vietnam, Laos PDR, and Thailand. Yu, Xue, Sun, and Zhang (2016) researched product transportation distance and environmental sustainability. The researcher suggests that firms should select sustainable suppliers for their products and also try to find nearest suppliers to reduce the carbon emissions and hazardous gases, which have a significant effect on the environmental sustainability and also increased the overall organizational cost with negative reputation and image in the global area. The supply chain offers integrated cost optimization of production, distribution, and sourcing through flexible suppliers' networks and manufacturers' operations. The reduction of environmental effects of freight inbound and outbound significantly increased sustainable development in the region.

Carbon emissions and greenhouse gases are the principal reason for bed environmental sustainability. The researchers suggest that firms should shift their polluted transportation towards green and clean transportation for good environmental sustainability with improved firms' image. The findings show that carbon management practices in transportation and supply chain can be observed only by the reduction of carbon emission activities and promote environmental sustainability through renewable energy sources. Davies, Thomas, and John (2013), the movement of goods is a necessary aspect of modern era global manufacturing systems. It is also recommended to reduce polluted products manufacturing and encourage green products market, while government and regulatory authority should play a vital role in coming in the front to control polluted activities through strict environmental-friendly laws and enforcement of environmental laws.

There is an increasing concern on emissions and cost on account of speed reduction for international shipping (Dongmin, Dingyi, Youfang, & Bin, 2016). Low steaming encountered these more enormous challenges of technical requirements and time constraints, while the emissions and cost of railway transportation distinctly differentiate from international shipping. Furthermore, senior management (IEM) may enforce environmental-friendly approaches and methodologies in the firms' supply chain management for the future betterment of environment, society, and firms' itself. Marchet, Melacini, and Perotti (2013) many types of research have addressed environmental problems in their findings and also suggested for improvement from the senior management and governmental regulations although researchers have started to check the process of decision-making involved in environmental development and sustainability adoption in the last couple of years, many themes, such as sustainability measurement and evaluation, and the level of implementation of sustainability initiatives in the context of 3PL (third-party logistics) industry.

The transportation sector, as the goods and passenger carrier, is doubtfully one of the basic infrastructures, which is necessary for industrial and economic growth. On the other hand, the transportation sector also emits a large portion of emissions, which have significantly and negatively effect on environmental sustainability, people's health, and firms' reputation. "Green transportation" is a big field, and different countries' researchers are working on this topic to promote green and renewable transportation modes, which is safe to environmental sustainability for the long term. The mode of road transportation and freight is a pivotal contributor to carbon dioxide emissions. Minimizing these carbon and GHG emissions in transportation route planning offers an understating of vehicle emission models and their effect on environmental sustainability.

The logistics with green practices have emerged as a trend in the distribution management of goods and collection after shelving life products, which enhanced the environmental sustainability and created responsibility for the consumers in terms of returned products after product life cycle. In many European countries, firms are already working on this approach to re-manufactured, recycle, refill, and reuse the products for greater environmental sustainability. On the other hand, these green approaches enhanced corporate image and reputation in the global market but also created significant financial benefits and improved environmental performance. In the last couple of decades, green product ideology is also introduced in Asian countries such as Pakistan, India, and Sri Lanka. Many firms of these countries are implementing green practices in their supply chain operations for long-term survival.

Buses and cars are the most used vehicles for urban public transportation mode. The contribution significantly increased to the pollution and health diseases in urban areas, while industrialization is also a significant cause of environmental degradation and health diseases, which also cannot be ignored. Researchers suggested the usage of renewable and clean energy sources in transportation mode to control global warming and temperature. Y. Geng et al. (2013) researchers researched the region of Shenyang, China. The findings show that CNG buses and cars have the best overall environmental and economic performance. On the other hand, petrol and diesel vehicles are not environmental-friendly and also increased the cost. Electric vehicles, plus higher costs, embodies emissions are more significant, making them more sophisticated as a public mode of transportation. To further improvement in the public transportation industry, integrated efforts are required, including the gradual phaseout of inefficient vehicles, infrastructure improvement, capacity building, and green vehicle purchase initiatives.

5.1.5 IEM and ED

The finding shows that internal environmental management has a direct and positive effect on ecological design, while a 1% increase in internal environmental management will lead to 0.50% improvement in ecological design on a 1% confidence level. Senior management, as a very active role in the adoption of green practices. On the other hand, governmental authorities and customer pressure also have a significant contribution, which forces firms to adopt the ecological design for environmental sustainability and reducing global warming issues. The development of green supply chain practices in firms is still challenging. Using an approach of theory building, researchers built a theory of social-ecological intergradation to offer senior management with guidance in mimicking ecosystems to build more local supply chains. The research paper identifies key five fundamental principles that mimic the natural system of ecology and also increase environmental sustainability. The article further draws what these fundamental principles mean in both a supply chain and an environmental context. Firms increasingly adopt the product design of eco-friendly techniques to lessen their harmful effects on environmental sustainability. Undeniably, the theory of ecological design is still in the early development stages.

The billion trees project started in 2013 by the central and provincial governments of Pakistan, which has been successfully ongoing, and due to this project, environmental sustainability significantly has been improved, and firms are successfully adopting ecological design in their products. On the other hand, many international organizations, institutions, and NGOs appreciated the Pakistan government project. Lai, Wong, and Cheng (2012) investigated how many ecological modernization forces in order of environmental regulatory laws, economic pressure, and customer pressure are correlated with the adoption of green logistics management by Chinese export firms to make the life cycle of logistics and their products. Researchers collected 128 firms' data from China, while the results show that the pressure of customers is a crucial factor affecting the extent of their ecological design adoption and green supply chain practices, which in turn is positively correlated with their operational, financial, and environmental performance.

In Pakistan, environmental friendly supply chains become more profitable, and several companies are switching their supply chain over the green or environmental friendly supply chain. However, the most critical question is why companies are switching over an environmentally friendly supply chain. So the answer to this question is a little complex, because there are several factors that are influencing companies to switch over the green supply chain and also many drivers are involved, creating pressure on management to implement an environmentally friendly supply chain. On the other hand, some barriers also exist which hinder firms from being environmentally friendly, while this research has uncovered the critical significant barriers and drivers involved in the phase of green supply chain implementation in the context of Pakistan. De Medeiros, Ribeiro, and Cortimiglia (2016) researched purchasing green products in Brazil. This article examines experimental research that the association between perceived value by the consumer of green products and price elasticity. The research results show that the perceived value of ecological products increases willingness to pay in the decision of purchasing. Besides, the demographic variables such as age, revenue, and gender have positively affected perceived consequent purchase intention. Results also show that the impact of green design is similar to the 10% price minimization. In simple words, the declared preference, respondents were able to pay. The fundamental theoretical contribution of this study lies in understanding the phenomena of how key dimensions constitute perceived value.

Industrialization leads to massive employment generation needs and also economic growth. However, firms also consider the environmental effect of industrialization to control their harmful effects on environmental sustainability and health concerns. In this regard, GSCM emerged as a strategy of an environment that enhances the performance of the environment but also improves firms' image and reputation with more significant profitability opportunities. Finally, the researcher suggests that senior management commitment is significantly crucial for the adoption of ecological design in supply chain management.

The idea of corporate responsibility is a little complex because most firms are adopting CSR (corporate social responsibility) to enhance their firms' financial performance. Nevertheless, the essence of this ideology is to improve environmental sustainability; some implications of the sustainability framework for practitioners include the effects of incorporating sustainability into external communications and corporate strategy to stakeholders. The definition of ecological design is provided by Khan and Dong (2017b) ecological design offers that procedures design products which minimize consumption of material and energy, which also facilitate the procedures of recycling, reuse, and recovery of materials and meanwhile minimize the use of harmful materials in manufacturing products. According to Kleindorfer et al. (2005), the first-mover advantage is where firms introduce eco-design innovations in their product design and increase advantages, including developing unique capabilities of manufacturing and getting royalties for green technology licensing. Ecological design product has an opponent's effect on the enterprise performance and also plays a significant role to build environmental-friendly competitive advantage.

Many companies are using eco-friendly practices into their product innovation in terms of achieving competitive advantage. Besides, green products are also facilitating in re-manufacturing, reusability, recyclability, and disassembly of products. The product of eco-design may significantly reduce the harmful effects on the environment. Ecological design products also enhance firms' performance in terms of higher financial and environmental performance with the significant building of competitive advantage. According to Khan and Dong (2017b), the results indicate that ecological items have a strong positive association with firms' environmental and economic performance. Furthermore, ecological design products minimize the products' costs and enhance the value of products.

Most companies introduce ecological design in their products to attain financial benefits, including receiving royalties licensing, which could provide a green competitive advantage. Product development and design processes may assess the ecological effects of goods until its whole product shelves life stage. Ecological design enhances financial performance by reducing the adverse effects of production operations in terms of recycling and re-manufacturing approaches. The adoption of green practices in the supply chain has a positive effect on the firms' financial performance, but sometimes green practices may negatively affect the firms' financial performance due to the wrong implementation of practices and without a detailed understanding of organizational and product strategy. Firms also try to adopt expensive green technology without a proper understanding of their supply chain strategy and their pros and cons on the financial performance, which may result in a drain on firms' profitability (Khan and Qianli, 2016).

Undeniably, the process of green supply chain management can be varied by industry to industry, company to company, and product to product. However, the bottom line of the green supply chain procedure is the same. Khan and Qianli (2016) conducted research on the effect of the green practice on the firms' performance. Researchers collected the sample of 415 manufacturing firms' data, and the green practices were measured by 5 critical factors, including green purchasing, green information systems, green distribution, and green manufacturing. Researchers used linear multiple regression analysis for testing hypotheses. The findings show that green design and/or eco-friendly product design has a significant effect on the firms' performance. Further, the researcher suggests that internal environmental management has an active role in the implementation of green practices, including ecological design in the overall supply chain management, because senior management has the authority to build organizational and supply chain strategy. The green supply chain management not only saves the environment. However, they also provide an opportunity for companies to grasp the competitive edge in their market and increase their profitability.

5.1.6 GIS and GM

The findings show that green information systems have a strong positive relationship with green manufacturing. In simple words, green information systems improve to green manufacturing in the organization, while a 1% increase in the internal environmental management will improve 0.34% to green manufacturing on the level of 1% confidence level. Green manufacturing can directly benefit the environment. For example, green manufacturing can help to reduce waste and harmful emissions and work towards preserving resources that are finite and nonrenewable. Many customers want to support businesses that implement green manufacturing, so by implementing this process, a business can also gain new customers. Because of the public's consensus about the need to protect the environment, a company can directly improve its public relations by implementing green manufacturing. Additionally, this process can lower costs for the business over the long term through the implementation of more efficient systems and fostering a company culture dedicated to innovation in processes. Further, these more efficient processes which can result in lowering the amount of waste a business produces.

"Going green" has become a more popular trend as of late, with more and more businesses opting to make the switch. While it can certainly seem overwhelming, manufacturers who embrace sustainable practices can recognize a positive impact on their bottom line. These benefits can help boost sales, lower operating costs, and more.

A manufacturing business can see overall operating costs drop if the green changes are planned accordingly. Having access to solar and wind energy, along with appropriate energy-efficient equipment and machinery, can drastically reduce energy and utility bills. Additionally, increasing recycling and going paperless for bills/documents can also save on supply costs. Similarly Khan and Dong (2017a) conducted the research and collected the 218 firms' data to test the association between the green information systems and their effect on the green practices and performance of the organization; the results indicate that green information systems enhance to the green practices and firms performance.

The idea of environmental-friendly manufacturing has garnered considerable attention worldwide. Several firms are implementing green manufacturing in their business operations to gain fruitful benefits of green practices.

Green manufacturing contains all elements correlated with ecological design in manufacturing by continuously integrating eco-friendly processes and items. The green manufacturing may mitigate water, air, and land pollution, minimize waste at the source, and reduce the health risk. Undeniably, the aim of green manufacturing may be achieve through proper design of products and manufacturing processes. The design of green 46.6% is the primary dimension of concern. Other dimensions are sorted to reduce the order of weight, while 34.4% is the green manufacturing and green packaging. The design of green products accounts for nearly 50% of overall weight, which can demonstrate that it is essential when firms adopt green manufacturing. The broadly used green practices in the automobile sector include green warehousing, green distribution, green manufacturing, and green packaging. The report shows that some environmental-friendly have a significant role in the reduction of carbon and GHG emissions (Wang, Subramanian, Gunasekaran, Abdulrahman, & Liu, 2015).

Many developed countries' manufacturers generally adopt green practices and have high levels of environmental concerns. On the other hand, green practices trend is new in the Asian market. However, in the last couple of decades, firms are continuously adopting green practices in their supply chain operations such as green manufacturing, green information systems, green purchasing-procurement, green logistics, and reverse logistics practices to enhance environmental sustainability and also create competitive advantage in their consumer market. Scur and Barbosa (2017) researched in Brazil, and researchers collected the data from different manufacturing firms to test their hypothesis. The results show that green information systems are the focal point of green supply chain management; firms are unable to implement green supply chain practices successfully without the adoption of green information systems because green information systems informed the firms regarding their previous, present, and future forecasted performance.

Mainly the green supply chain management is based on green information systems, green manufacturing, green purchasing, green marketing, and services; while these practices cover green competencies, the amount of resource utilization and green management abilities make good use of ICT tools, efficiency-based strategy, closed-loop strategy, risk-based strategy, and innovation-based strategy. Liu et al. (2012) researchers surveyed the Central China region and collected the data of manufacturing firms, while after using factor analysis techniques and testing hypotheses, the results show that green manufacturing, green distribution, and green logistics have a significant effect on organizational performance. However, the critical role played by green information systems, which measure the firms' performance and inform to the updated schedule and green plan to the management. Researchers also confirmed that firms' competitive advantage also might be built due to the adoption of green manufacturing in their production operations.

Khan and Dong (2017a) the impact of firms' sustainable activities on firms' performance, while researchers collected the data from different 415 firms and used the step-wise regression analysis to test the hypothesis. The findings indicate that green information systems, green manufacturing green purchasing, and distribution have a positive and significant effect on the firms' financial and environmental performance. However, the key findings were green information systems which enhance and monitor the implementation of other green practices. Researchers also suggested that future researchers conduct further study on green transportation systems and green logistics effects with different green practices on organizational performance.

5.1.7 GIS and GD

The green information systems and green distribution both have a strong positive correlation on the 1% confidence level, while a 1% improvement in the green information systems will bring 0.47% enhancement in the green distribution systems. In simple words, green information systems lead to green distribution, and both have a significant and positive relationship.

Sustainable distribution practices include those that reduce carbon dioxide, are economically viable, and will bring about a better quality of life for the earth's future inhabitants. Green distribution practices range from changing the way distribution centers and vehicles are powered to implementing greater transparency regarding the environment and distribution practices. As environmental concerns increase, the integration of environmental issues into the supply chain studies has become a thriving subfield.

The storage facility is another essential aspect of green distribution. The storage facility should be able to store different categories of materials. Businesses should use sustainable practices when powering their storage facilities. Instead of powering these storage facilities with electricity, companies can use hydro, wind, or solar power, further optimizing resources. Furthermore, the design and construction of storage facilities should meet the prerequisites of a non-polluted environment while strengthening the maintenance of proper humidity, corrosion, and waterproofing in addition to other factors. This will not only help generate efficiencies but will also enable organizations to contain costs. As organizations restructure to reduce their company's environmental footprint, supply chains have increasingly become a key area of focus. Improvements in transportation efficiency and packaging should be at the top of the green supply chain initiatives list (Liu et al., 2012). The research paper examines the association between green supply chain management practices and their effect on the competitive edge and organizational performance. Researchers used factor analysis and structural equation modeling techniques to test the hypothesis. The results show that green distribution has a strong effect on the firms' performance, while green information systems are playing and supporting a significant role in the implementation of green practices. Furthermore, green distribution also creates a competitive advantage for the firms in the international and local markets.

The green distribution consists of green packaging. Packaging features such as size and materials used influence distribution because of their effect on the transportation features of the product. Green packaging involves downsized packaging and the use of green packaging materials. They also point out the need to cooperate with vendors to standardize packaging, encourage and adopt returnable packaging methods, and promote recycling and reuse of packaging materials. Khan, Dong Qianli, and Zhang (2017) researched the green practices in the supply chain and their effect on the firms' performance. The green practices variable is covered by green distribution green information systems, green manufacturing, ecological design, and green purchasing. The results show that green distribution has a strong correlation with green information systems and green manufacturing and these significant effects on the firms' performance. Furthermore, researchers suggested that the top management role is also very significant to enhance adoption of the green practice in business operations because senior management has the authority to change organizational and supply chain policies towards green.

Green information systems and green supply chain management are two strategies for internal and external orientations for the green development of a company. This research recognizes the prevailing direction between the two information technologies enabled innovations in order of process control, environment protection, and firm support. Green information systems and green supply chain fit are based on both fit-as-moderation and fit-as-profile-deviation approaches to capture the synergy of green practices (Yang et al., 2013). The capability to share information with SC members, coupled with the specific capabilities of GIS, improves the performance of the environment. The GIS serves as a partial mediator to the association among SC environmental performance and sharing information.

Several challenges are associated with green distribution such as the high start-up investments associated with employees training and environmental auditing, limited financial resources, the attitude that firm's strong environmental commitment results in added costs which place the firm at an economic disadvantage as compared with other less environmentally responsible firms, and reduction of the pool of qualified suppliers due to stricter environmental standards. Khan and Dong (2017a) researched green supply chain practices effect on the firms' performance in the context of Karachi. The researchers collected the data of around 218 companies from different industries. The green supply chain was measured through five key independent factors including green purchasing, green information systems, green manufacturing, cooperation with customers, and eco-design after applying exploratory factor analysis and regression analysis for testing hypothesis. The research results show that factors have significantly and positively affect organizational performance, including green manufacturing, green information systems, and ecological design. Further researchers suggested that senior management should not only adopt ecological design in their supply chain but also concentrate on the proper implementation of GIS (green information systems) because green information systems monitor the firms' performance and provide future forecasts and schedules for the management/plan.

The distribution function of an organization aims to address the customer-organization relationship issue. For instance, every market has to be informed about the products manufactured, their features, their availability, and their various capabilities. Besides, manufactured goods have to reach the market on time. This necessitates suitable distribution systems. Environmental concerns are considered by offering environmentally friendly products through environmentally friendly distribution systems. Green distribution is achievable through green packaging, green warehousing, and green storage. In order for firms to fully implement and realize potential gains of green distribution, manufacturers should comprehend the requirements and ensure that their products conform to the environmental concerns in addition to controlling and accessing their distribution patterns to enhance the financial and environmental performance of the firm. According to Woo, Kim, Chung, and Rho (2016), GIS is essential for enhancing and cooperation among supply chain holders through information sharing. The findings indicate that green information has a positive and significant effect on green practices such as green distribution, green manufacturing, and ecological design, which finally create a positive effect on organizational performance. Green et al. (2012) researched the sustainable supply chain and their effect on enterprise performance and manufacturing operations. The adoption of green supply chain management practices by manufacturing firms achieves enhanced economic and environmental performance, which also has a significant positive effect on the operational performance of the firms. Green information systems and internal environmental management are recognized as compulsory precursors to the adoption of green practices including investment recovery, cooperation with customers, green purchasing, and ecological design.

During the implementation of green supply chain best practices, a green information system plays a vital role in sharing reliable information with suppliers and downstream levels. The green information system creates active collaboration and coordination among and within supply chain departments. The dream of green supply cannot become real without the proper adoption of green practices in the firm's distribution, while the distribution operation is more connected with the downstream level. Green et al. (2012) researched environmental collaboration and improved supply chain performance. This article aims to contribute to the first wave of empirical investigation regarding the effect of GSCM practices on firms' performance and their environmental performance from the manufacturers' perspective. Researchers collected the data of 159 firms of the USA and used the structural equation modeling to test the hypothesis. The results indicate that environmental monitoring practices among supply chain members are found to achieve an enhanced overall firm's performance. To achieve the critical objective and get fruitful benefits of green supply chain management, firms should adopt green distribution and green information. These two factors are playing significantly in the successful overall implementation of green supply chain project. In the perspective of Pakistan, in the last couple of decades, the government have changed their environmental policies, which are the significant reason for the growth in the green supply chain trend. On the other hand, firms also adopt green practices for their operations improvement and financial and environmental benefits.

The findings from this study clearly show that green distribution practices significantly and positively influence firm competitiveness, improve financial and environmental performance, reduce cost, and improve efficiency. To be able to curb waste in Pakistan, manufacturing firms should be encouraged to embrace more of the green distribution practices. This will enhance environmental sustainability. Green distribution practices are thus a good strategy which the management should be committed to in order to remain innovative, effective, competitive, and efficient in today's ever-changing dynamic marketing environment.

5.1.8 GIS and GT

The results show that green information systems and green transportation have a positive and significant relationship, while a 1% increase in green information systems will lead 0.36% increase in green transportation on the level of 1% confidence level. In simple words, these green information systems and green transportation have a significant role in the adoption of green supply chain management into firms' operations. The transportation sector emits heavy pollution, including carbon and GHG, which significantly destroy environmental sustainability and also create health problems. Many studies have been conducted on environmental sustainability and their effect in European countries, but still, there is no such in-depth research conducted in Pakistan, which can find the causes and barriers face during the adoption of green practices. The green practices implemented in business operations are not only protecting the environment but also bringing positive change in the financial performance of the firms. It is also a key reason for implementing the environmental supply chain. Many types of research have been conducted to prove that firms adopting green practices in the supply chain will significantly improve financial performance.

Consumer pressure playing a significant role in encouraging firms to adopt green practices in the end-to-end supply chain. Firms manufactured products for consumers, and if consumers are not willing to buy products, so in this situation, firms cannot survive in the long term. In today's world, consumers are very aware of every information; the Internet also plays a critical role, and consumers can get all information in a click.

In the green supply chain projects, firms cannot ignore to the adoption of green practices in the transportation, undeniably green transportation play a significant role, and green practices are unable to deploy in transportation without the top leadership commitment towards green projects. Transportation emits massive carbon dioxide and greenhouse gases, which create a significant and negative effect on environmental sustainability in the long term. Rostamzadeh, Govindan, Esmaeili, & Sabaghi (2015) confirmed that transportation emits 23% of carbon emissions and 15% GHG (greenhouse gases), which are the cause of climate change and global temperature. Herold and Lee (2017) this article's objective is to identify the green practices and their importance in the supply chain management, researchers collected the data from logistics and transportation sectors from 2000 to 2015. The results show that the transportation industry is a cause of higher carbon and GHG emissions, which not only causes environmental degradation but also creates severe diseases in the region.

The IEM (internal environmental management) represents the senior leadership commitment towards green supply chain management projects. The senior management cannot ignore the transportation movement and negative effect on the environment if they have decided to adopt green practices in the end-to-end supply chain management. Generally, internal environmental management emphasizes the adoption of green practices, namely, green manufacturing, green design, green distribution, and green transportation. Green transportation has a significant and positive correlation with a competitive edge and environmental performance, which are the bigger goal of the firm in the implementation of green practices. Usually, in the green transportation, firms using eco-efficient fleet such as energy-efficient vessels and using green fuels like low sulfur content, proper maintenance of the vehicle, provide training to the employees and achieve green transportation certification from government/regulatory authority, and these are the key points firms need to implement in green transportation.

On the other hand, green transportation not only increased environmental sustainability but also enhances and develops a competitive edge, firm image, and reputation, and most times government provides subsidies and tax exemption to the firm. According to Khan and Dong (2017a), green transportation may reduce environmental pollution; researchers researched the context of the Asian region. The finding shows that green information systems are the critical factor, which forces management to adopt green transportation and other green practices in the supply chain. Researchers used a linear regression method to test their hypothesis. The findings show that green purchasing has a negative association with organizational performance.

On the other hand, green information, cooperation with customers, green ecological design, and green manufacturing have a positive and significant effect on the firms' performance in terms of financial and environmental. Further, researchers also suggested that managerial staff should adopt green practices after an in-depth understanding of their advantages and disadvantages. Most times, management procures massive green information technology, which has no such compatibility with the firms' systems while decreasing the overall profit of the firms.

Environmental-friendly transport refers to the adoption of green practices in transportation to reduce the harmful effect of the transportation operation on environmental sustainability and climate change. The movement of goods and bought-out parts is an essential aspect of modern-day worldwide manufacturing systems. Many types of research have addressed the problem of global warming and environmental challenges within the supply chain. Kirschstein and Meisel (2015) researched environmental sustainability issues and their solutions. Intermodal transportation is an instrument of green logistics, which may significantly reduce GHG and carbon emissions. In order to assess the environmental impact of rail transport and road transport, researchers have formulated very detailed microscopic models, which examine emissions of vehicle precisely based on a vast number of parameters. One of the objectives of this research paper is to build models that consolidated the preciseness of models while offering only little more information than models of the macro. Besides contrasting research papers considering either rail mode or road mode, we offer models frequently for both transportation modes. We found that emissions of transportation may also be controlled through the usage of renewable and clean energy, and also firms try to find the nearest supplier, which may significantly reduce the distances/lead time between origins and destinations.

Transportation has appeared as a critical factor behind the emission of harmful pollutants, including carbon emissions, nitrogen oxide, and particulate matter (PM2.5). The aim of green transportation is to reduce fossil fuel consumption and enhance or build a firm's competitive edge in the international arena. Liu et al. (2012) conducted a study on the impact of green supply chain practices on firms' competitive advantage and performance from the perspective of China. Researchers collected the data from different manufacturing firms of China and used structural equation modeling on LISREL software to test the hypothesis. The findings show that green information systems and green transportation have a strong positive correlation, and both practices are encouraging firms to adopt green supply chain management in their business operations. Furthermore, green transportation also creates a competitive advantage for the firms in the international arena with more significant firms' financial and environmental performance.

This research paper examines the effect of five different determinants of the GSCM practices on enterprise performance in the context of Pakistan manufacturing firms Khan and Dong (2017a). Researchers collected the data from manufacturing firms. The GSCM practices are measured through four variables, including cooperation with customers, ecological design, green purchasing, and green information systems. By using exploratory factor analysis and linear regression method, the findings show that except green procurement, remaining four variables have been found positive and significant to forecast firms' performance. However, the ecological design of green practices followed by GIS has shown the most significant impact on firms' performance. Therefore, the top management of the firms should not only adopt ecological design in their business operations but also concentrate on proper monitoring of GIS to enhance firms' performance. The critical contribution of this research from the theoretical side is that it is possible to notice a negative effect of green procurement towards firms' performance principally in the context of the Pakistani manufacturing sector. One more key finding is that green procurement is a significant antecedent of companies' financial performance in the US manufacturing companies although not mainly correlated to firms' performance in our research. Furthermore, top management commitment is also essential in the overall implementation of green supply chain management project.

Generally, a green information system is developed, if the senior leadership is well-committed towards the green supply chain project and green IS increased to the information, monitoring of the green project and also focused on environmental performance evaluation. The primary objective of the green supply chain is to improve environmental sustainability and enhance economic benefits and create a competitive edge with positive image building in the customer mind. On the other hand, green transportation is mainly working on the reduction of carbon emissions, fossil fuel consumption, and increased renewable energy sources to promote environmental sustainability. In the last couple of years, researchers pay attention to the role transportation plays in achieving positive environmental performance. Green transportation can smash with the designated financial goals. Exploring the association between environmental effects and transportation by the planning of route will be able to offer valuable and practical suggestions regarding green supply chain management. The vehicle with advanced green technology, such as using more aerodynamic trucks and more fuel-efficient/fuel-powered trucks, is crucially recommended for the green supply chain projects.

5.1.9 GIS and ED

The results indicate that green information systems and ecological design have a positive and significant association on a 1% confidence level. Besides, a 1% improvement in green information systems leads 0.27% improvement in ecological design. In other words, green information systems create a positive effect on ecological design, which also enhances firms' overall profitability in terms of financial performance and environmental performance. Ecological design plays a vital role in the adoption of green supply chain management projects, while the green information system is the motivating factor of ecological design. Product stewardship incorporates in the situation of external members into the ecological development stages through product life cycle analysis; green products can measure the product's effect on environmental sustainability. Firms also build a competitive edge through the adoption of ecological design and enhance environmental and economic benefits and also develop unique manufacturing capabilities. Ecological product has a significant and positive effect on the firm's survival and plays a significant role of weapon to create a sustainable competitive edge in the international market.

Undeniably, green design and environmental-friendly practices are mainly dependent on customer requirements, supplier and customer management, and senior leadership commitment. Firms may reduce more than 80% of their adverse and harmful effects from their green products and processes on environmental sustainability due to the successful adoption of ecological practices in the product design phase. Besides, ecological design may reduce the usage of toxic and hazardous chemicals in their product and also minimize resource consumption in production goods. In similar Nie (2016) researched green supply chain management and ecological design. Researcher results indicate that ecological design has a strong effect on firms' performance, while due to ecological design, companies can reuse, recycling, and re-manufacturing approaches also used. Furthermore, research shows that the adoption of the green supply chain will not only enhance economic and environmental performance but also create a competitive advantage for the firms.

Ecological design facilitates firms in the re-manufacturing, reusability, recyclability, and disassembly of products, while green products have a direct effect on firms' environmental and economic performance. Customers prefer to buy green products to attain the benefits of environmental protection, and firms reduced their cost through the adoption of green design approaches. Also, firms build a competitive edge, and the implementation of green design practices significantly enhanced company image and reputation in the international arena. Green product design reduces the cost of goods and mainly improved the value of the product. According to Liu et al. (2012), green information systems and ecological design play a vital role in the implementation of green supply chain management techniques and practices. Researchers used factor analysis and structural equation modeling to test their research's hypotheses. The findings show that green manufacturing, green distribution, and ecological design have a positive effect on the firms' performance in terms of financial and environmental. However, green information systems are the initial stage, which monitor the success and implementation of green supply chain management.

Khan et al. (2017) researched sustainable supply chain practices and their effect on the firms' performance. Researchers collected a sample of companies from different manufacturing firms. Mainly the green supply chain practices were measured by five factors, including green manufacturing, green distribution, green purchasing, green information, and ecological design. Researchers used regression analysis for measuring their research's hypothesis. The results indicate that green distribution has no significant effect on the firms' performance. On the other hand, green information systems, green manufacturing, and ecological design have a direct and significant positive effect on the firms' performance. Furthermore, researchers suggested that the management staff of the enterprises should adopt green practices in their business operations to promote green practices and also enhance their firms' financial and environmental performance with competitive advantage building.

Khan and Dong (2017a) researched the green practices in the manufacturing firms of Pakistan and their impact on firms' financial profitability. Researchers collected the data from different companies and used exploratory factor analysis with simultaneous regression analysis to test the research's hypothesis. The findings show that green purchasing is negatively correlated with firms' performance, while ecological design and green manufacturing systems are firmly and positively affecting the firms' performance. On the other hand, researchers found that green information systems have a vital role during the adoption of green practices in supply chain management. GSCM and GIS are two strategic endeavors of internal and external orientations for the green development of a firm, while the results confirm that green information systems are used for monitoring of green supply chain project, and it also helps management to inform regarding the weakness of the project. On the other hand, green information systems and ecological design create positive effect over organizational performance, and ecological design also helps firms to reduce their cost of the supply chain in terms of recycling, reuse, re-manufacturing, and refilling approaches.

GIS offers the information needed for collaboration and coordination with downstream and upstream in terms of green design, labeling, and packaging. Information sharing, through the use of GIS, is the primary enabler for supply chain management/sustainable supply chain management in order of integration, collaboration, and coordination. GIS also provides the necessary information, which helps to make decisions about green design in terms of energy and material consumption, recycling, reuse, recovery of materials, and re-manufacturing of products. Besides, GIS offers the information required to recover the firm's investment in scrap, excess inventories of material, and finished goods.

5.1.10 GM and ECP

The findings show that green manufacturing and economic performance have a statistically positive and significant relationship. Green manufacturing encourages firms to reduce their unnecessary processes, waste and clean their processes, which directly affects the firms' profitability, while a 1% enhancement in green manufacturing processes will significantly increase economic performance by 0.30%. Green supply chain management projects main idea is to implement or adopt green practices in the firm to reduce carbon emissions, climate change, and global warming while increasing environmental and economic performance with the higher customer service level. Several firms are adopting green practices to improve their financial statement and image in the international market and also grab maximum export opportunities. In general speaking, green manufacturing increased the efficiency of using raw materials and components with the least consumption of hazardous materials and toxic chemicals. Besides, increased environmental sustainability reduces environmental pollution such as carbon emissions, GHG (greenhouse gases), and global warming issues.

Green manufacturing processes are very flexible and also supportive of re-manufacturing, recycling, and refurbishment of products after PLC (product life cycle) or PSL (product shelve life). The firms have the plan to implement green projects in their end-to-end supply chain management need to mainly focus on green manufacturing practices, which are called the heart of GSCM (green supply chain management). These firms conduct preventive maintenance and predictive maintenance, emphasizing on their product performance and enhancing their PLC; they also conduct disassembly of the finished goods for subsequent reuse/re-manufacturing/recycling (Scur and Barbosa, 2017b). Greens et al. (2012) researched green supply chain practices and their impact on firms' performance from the perspective of the USA. The data is collected from 159 different manufacturing managerial staff and is used as structural equation modeling method to test the hypothesis. Managers of manufacturing offer data reflecting the degree to which their firms work with customers and suppliers to enhance the environmental conditions of the supply chain. The findings show that the adoption of green supply chain management practices by manufacturing companies significantly achieves enhanced financial and environmental performance, which also turns into a better competitive advantage and operational performance.

Green manufacturing mainly covers material exploitation, production planning, control activities, waste reduction in terms of material and energy consumption, processing time, waiting time, setup time, etc. and also helps firm to achieve their financial goals. Pakistan consumers are also willing to buy green products, and that is why Pakistani manufacturers are now focusing on GSCM and especially emphasize implementing green manufacturing as a pilot project. However, there are electronic firms, which are not willing to cuddle GSCM efficiently and effectively, as they already have long-term relationship with the targeted customer base and required no green manufacturing over to win or retain their customers. According to Khan and Dong (2017a), investment in cleaner technologies improved to environmental sustainability and performance. However, cleaner technologies also built a competitive advantage and improved the financial performance of the firms. Many firms are investing in implementing green practices, including Procter & Gamble, Engro Corporation, Unilever, etc. Zhu and Sarkis (2007) researched green initiatives and their effect on the economic performance of firms in the context of China. The evidence indicates no significant improvement in financial performance, while minor enhancement is measured in the environmental performance and operational performance of firms.

The manufacturing of green covers best resources, which, in the long term, may achieve a competitive advantage and build a positive image for the firms through a higher level of quality products at a reasonable price. Lean and green manufacturing works for the elimination of waste, increasing the efficiency of manufacturing with fewer defects along with an improved company image. In European countries have integrated and TQM philosophy, green and lean manufacturing practices to grasp higher environmental performance. Wee and Quazi (2005) researched to find the association between enterprise financial performance and green supply chain management approach. The results show that there is no such convincible relationship between financial performance and green supply chain management methods. Greening different operations in the supply chain will indirectly lead to the financial and environmental performance. Tseng and Chiu (2013) argue that manufacturing with green practices plays a vital role in implementation green supply chain project to eliminate adverse effects of manufacturing operations with little waste in overall production systems and also increase firms' efficiency.

Khan and Dong (2017a) conducted a research study on greening products and the impact of green supply chain management on organizational performance. Researchers collected the data from 218 companies, and the GSC practices were measured through five independent variables, including green purchasing, cooperation with customers, green manufacturing, ecological design, and green information systems, while researchers used different statistical methods with exploratory factor and regression analysis. The results indicate that green manufacturing, ecological design, cooperation with customers, and green information systems have a direct and positive association with organizational performance. On the other hand, green purchasing is negatively correlated with firms' performance, while due to less and/or no governmental and regulatory authorities' incentives, mostly time green raw materials and components are costly, which is a burden on firm's financial position without the incentives from the governmental side. Schniederjans and Hales (2016) researched green supply chain cloud computing and its effect on the economic and environmental performance of the firms. Companies are continuously facing pressure from supply chain members' side to adopt green practices in business operations. Undeniably, economic performance is a significant driver for firms that wish to adopt environmental initiatives, because economic problems and challenges related to costs are usually the key barriers and hurdles to adopting green initiatives in the supply chain management. Thus, we emphasize on balancing environmental performance and economic performance. The definition of economic performance interchangeably uses with the business performance or firms' performance. Many firms throughout the globe face regulatory and customer pressures to adopt green practices in their supply chain management for better survival.

Many studies founded a positive association between economic performance and green practices, while green practices enhance environmental performance for the firms' but also increase the economic performance of the firms including quality and cost performance, stock market performance, consumer market performance, innovations, and manufacturing performance (Schniederjans & Hales, 2016). To a slighter degree, a couple of studies have also assessed critical factors of success for enhancing environmental performance of the firms including, sales and export to foreign consumers, manufacturing process improvement, product design improvement, and interfirm linkages facilitated by proximity demands)and cost-based (dimensions relating to organizations reducing their cost structures) measures. Demand-based measurements include a focus on the reduction of solid waste, consumption of hazardous/harmful and toxic materials, resource consumption, improvement in environmental reputation, and environmental certification (Hayami, Nakamura, & Nakamura, 2015).

Cost-based measurements include decreasing costs for materials purchasing, energy consumption, waste treatment, and waste discharge.

Firms minimized their waste and cost from the system by adopting green processes and technologies. Lean manufacturing and green manufacturing both are working, in the same way, to control and reduce waste from the manufacturing system and eliminate unnecessary processing steps and increase the efficiency of manufacturing along with an improved company image and reputation in the domestic and international arena – measuring the environmental performance of green practices (Schniederjans & Hales, 2016). Undeniably, environmental challenges are complex to quantify; comparing the impacts of environmental practices' impact on companies' performance can be problematic when the environmental quality data is not correct or inadequate. Khan et al. (2017) study explored the association and linkages between green practices and enterprise performance in the context of Chinese manufacturing firms. Researchers used predictors in their study, including green purchasing, green manufacturing, green distribution, and GIS (green information systems), an ecological design that was measured for the practices of green supply chain management. After the use of mandatory statistical analysis, researchers used factor and regression methods to test their research's hypothesis. The results show that green purchasing and manufacturing have a significant effect on enterprise performance. On the other hand, ecological design and green distribution are also affecting organizational performance (Khan et al., 2017).

Green manufacturing provides firms an opportunity to differentiate their services from other manufacturing firms and build a competitive edge. The results show that green manufacturing has a significant and positive effect on the firms' economic performance. Green manufacturing practices mainly reduced waste from the processes in terms of unnecessary processes, reduced processing time, and setup time. Companies are becoming more reliant on the vendors and key suppliers to achieve economic gain. Senior leadership is increasingly emphasized to deal with environmental and social issues to enhance financial performance.

5.1.11 GM and CA

In the research, results reflect that green manufacturing and competitive advantage have a statistically positive and significant association on the 5% confidence level. In other words, green manufacturing creates a competitive advantage for firms and companies to adopt green manufacturing practices to build their competitive edge in the domestic and international markets. Furthermore, a 1% increase in green manufacturing practices will lead to a 0.14% increase in competitive advantage. Companies might be held accountable not only for their individual but also for the social performance and environmental performance of their suppliers. The largest manufacturing companies could be characterized as marketing differentiators whose competitive edge is based on the product image and firm reputation. As a response to pressures from many supply chain members/stakeholders, such as customers, regulatory authority/government, NGOs (nongovernment organizations), and competitors, companies have introduced supplier evaluation schemes, which integrate social and environmental criteria. On the other side, there are many strategic drivers and motivations factors to adopt green supply chain management, such as improved corporate image, increased efficiency, increased innovation leadership, enhanced firm reputation in the international market, and also trust development. Proper implementation of green manufacturing practices may create innovations that lower the products' total cost or enhance its value in terms of customer satisfaction, which creates a competitive advantage in the market and also increased firm market share and size. Furthermore, green manufacturing is a suitable and appropriate approach to reduce environmental risk and enhanced corporate social responsibility. In similar line Liu et al. (2012) conducted the surveys in Central and South China market, researchers collected the data from manufacturing firms, and results show that green practices including green manufacturing, ecological design, and internal environmental management significantly create a competitive advantage for the firms in the long run. On the other hand, green purchasing has no such an effect on the substantial competitive advantage. Finally, researchers suggest that companies should adopt green manufacturing practices to increase their financial and environmental performance and also may reduce the overall system cost in terms of recycling, reuse, and re-manufacturing approaches and techniques.

The green manufacturing practices provide an opportunity for firms to create competitive advantage and also provide a chance to reduce their overall supply chain management cost in terms of recycling, re-manufacturing, and reuse. Container shipping on the basis has become mainly profitable and is especially significant for exporting firms to the global arena, which increases the competitive edge of the region even further through the 2030s till now. On the other hand, advances in technology and environmental-friendly policies have ensured to discourage firms from violating environmental laws and policies. Besides, environmental incidents may strengthen government and regulatory pressures, damage the company's reputation and image, and devaluate a firm's share price value in the stock exchange, and customers may stop buying/boycott the company's products and/or cancel their orders.

Firms are continuously facing massive pressure from consumer, society, and legal affair's pressure to implement environmental-friendly initiatives and practices to protect environmental sustainability and to reduce the harmful effect of business operations. Further, green practices offer an attractive opportunity for firms to increase their competitive advantage in the market and increase their market share and size through the adoption of green practices in their supply chain operations (Reimann, Schilke, & Thomas, 2009). Khan and Dong (2017a) researched green practices and supply chain performance; researchers collected the data of companies from the manufacturing industry of Pakistan. The green practices include green information systems, green manufacturing, green purchasing, and ecological design. The results show that green manufacturing has a significant effect on organizational performance, while companies may also reduce their overall cost through the adoption of green manufacturing techniques and practices in terms of recycling and re-manufacturing of products.

On the other hand, green information systems and ecological design help firms to adopt green manufacturing practices. Due to the adoption of green design, firms may easily recycle and re-manufacture their products (Khan & Qianli, 2017). Furthermore, researchers show that green purchasing negatively correlated with organizational performance due to low/or no incentives provided by regulatory authority side and green raw materials and components are much expensive than nongreen materials and components, which directly affect the profitability of the firms and mostly time negatively associated with firms' financial performance.

Khan et al., (2019), studied the relationship between green manufacturing and green supply chain practices' impact on environmental and competitive advantage. The results indicate that green manufacturing and ecological design have a significant and positive correlation with firms' environmental and financial performance, while green manufacturing also supports to firms in building a competitive edge and improving company image and reputation. Khan et al. (2017) researchers conducted a study on Chinese enterprise to identify the effect and relationship of green practices and organizational performance. Researchers collected the sample size of 415 manufacturing companies in Central and South China. After the testing of the research's hypothesis, researchers identified that green manufacturing has a significant effect on organizational performance, and green manufacturing also creates a competitive advantage for the firms. Senior management of the firms should adopt green practices for their survival in the global arena and also for environmental betterment.

5.1.12 GM and ENP

The results indicate that green manufacturing practices and environmental performance have a statistically positive association on a 1% confidence level. Green manufacturing supports the firm to adopt green supply chain management and also helps to reduce the overall system costs in terms of total quality management, lean manufacturing techniques, and waste elimination techniques. Further research results show that a 1% improvement in green manufacturing practices will achieve environmental performance by 0.22%. The findings show that environmental performance should be reviewed by senior management for further improvement, while monitoring system for data quality, green manufacturing practices quality, and evaluating systems also encourage firms to think over green supply chain management. Green manufacturing practices increase environmental sustainability in the region, which also plays a vital role in attracting foreign investors to invest in the country. Mainly the green manufacturing deals with the adoption of environmental-friendly means while implementing many actions in the progression of an SC (supply chain). The green supply chain management is an essential term that covers within itself a complete concept including green manufacturing, material management, green purchasing, green marketing, reverse logistics, green distribution, and waste management. In the traditional supply chain, manufacturing practices were using dangerous, hazardous, and toxic chemicals, which not only destroy environmental sustainability but also create a significant harmful effect on the flora, fauna, and human lives.

Green manufacturing minimize the harmful environmental effect by selecting reused and recycled products and/or goods which have been re-manufactured/recycled and refurbished, while green manufacturing practices create maximum flexibility in the manufacturing processes for recycling and re-manufacturing of products, which not only save money but also enhanced environmental sustainability in the long run. Green supply chain management theory is mainly based on green manufacturing and supply chain technology. The aim is to reduce the negative environmental effect and enhanced the fundamental objective of firm performance through the actions of the supply chain, management strategies, and collaborative partnership. Green supply chain management creates an opportunity for firms to enhance their overall firms' performance and image. Azevedo et al. (2011) suggest a model recognizes the influence of many green techniques on supply chain performance, while the researchers conclude that there is a positive association among green practices adoption such as customer satisfaction increased by adoption of green manufacturing practices and reduction in business waste; overall systems efficiency has been increased due to a reduction in waste and also a reduction in costs; firms create competitive advantage in their market; and governmental (Schenkel, Caniëls, Krikke, & Van Der Laan, 2015) authorities also offer incentives to the firms. Finally, researchers suggest that managers should adopt green practices to improve their economic, operational, and environmental performance with a more significant competitive advantage.

Many types of research identified the effect of green practices on firms' performance. Liotta, Stecca, & Kaihara (2015), the model was built in order to act as an efficient tool to understand the effect of manufacturing and transport on the environmental and economic performance of supply chain management. The efficient and effective coordination is a hallmark for betterment in green supply chain management; without such vendors and buyers, coordination will act independently to create a decision that optimizes their firms' goals. This non-coordinated approach is not successful in today's global world, due to fierce competition and customers' pressure on firms to adopt green practices, before competitions were between firms to firms, but now the trend has been changed; competition is between supply chains to supply chain of another firm. Carter et al. (2000) recognize the impact of green manufacturing and green purchasing on company environmental performance; they also found that successful adoption of green manufacturing activities enhance environmental performance and also build a competitive edge. The ecological design items have a significant and positive effect on firms' environmental performance. Besides, customers are more willing to pay for green products, while firms have more focused on green manufacturing and environmental sustainability to enhance their market share and size. Furthermore, governmental regulatory authorities also create pressure on firms to adopt green processes for manufacturing products, if any firm violates the laws of government, they directly enforce heavy financial penalties to discourage firms from adopting nongreen practices in the region. On the other hand, many countries around the globe are discouraging nongreen products from importing into their countries, which also create nontariff barriers for the polluted products/goods manufacturers.

Green manufacturing and lean manufacturing are working in the same direction to reduce waste and scrap and fewer manufacturing steps and increase the efficiency of the overall supply chain systems. The result showed that green manufacturing practices increase environmental performance (Tseng and Chiu, 2013). The green practices, including green manufacturing, reduce the harmful effect on environmental sustainability and enhance overall system efficiency with a positive reputation and image of the firms. Firms may adopt green supply chain practices in their business with the senior management commitment; mostly times senior management are not severe enough towards the implementation of green practices in their firms, which results in "failed green project or drain on firms' financial performance."

Green manufacturing is an ideology to control the overall supply chain cost and enhance the profitability of firms with green technologies, green processes, green material, and green approaches. Liu et al. (2012) Chinese firms and used the different statistical analyses to test their hypotheses, while results show that due to the adoption of green manufacturing practices environmental performance of the firms has been increased significantly. Because green manufacturing techniques are the group of total quality management and lean management, which eliminate the waste from the system and increase value addition. Khan et al. (2017), a study conducted on the green supply chain techniques and environmental performance of the firms. Researchers collected the data from different Chinese manufacturers and applied different statistical analysis with factor and regression analysis. The findings show that green supply chain management have significant and positive effect on the firms' performance, while due to adoption of green practices, firms also may enhanced their financial performance due to recycling, re-manufacturing, and reuse techniques, which not only reduce the harmful effect on the environment but also increase the environmental sustainability and reduce the supply chain cost. Further, researchers suggest that senior management should be well-committed towards green supply chain projects for successful implementation.

5.1.13 GD and CA

The findings of our research show that green distribution and competitive advantage have a significant positive correlation on the 1% confidence level, while a 1% improvement in green distribution will bring 0.33% betterment in competitive advantage. In simple words, green distribution and competitive advantage have a long-term relationship. However, due to the adoption of green distribution practices, firms quickly can gain a competitive advantage in the market. Furthermore, green distribution also supports firms in adopting end-to-end green supply chain practices in business operations. Green distribution practices create a positive image in the customers' market and fulfill the requirement of their customers to increased environmental sustainability, which creates a competitive edge for the firm survival in the long run. The results show that competitive edge is built by the adoption of green distribution in the end-to-end supply chain. On the other hand, green supply chain management is a broader name of green concept to enhanced environmental sustainability without creating a negative effect on the substantial financial benefits.

In the last couple of years, many companies are adopting green practices in their supply chain to improve the firm's performance, and companies have understood the criticality of environmental-friendly distribution network importance. Nevertheless, the starting cost and time investment to convert to this kind of distribution center terrify various companies. To enhanced corporate social responsibility and reduced distribution costs, many distributions and fulfillment firms are looking to adopt green practices that reduce their carbon dioxide, carbon footprint, and environmental pollution. The following criteria are mainly used to adopt green practices in the distribution network: green packaging and labeling, inventory level reductions, investment recovery, sale of unnecessary materials, excess inventory and scrap, etc. In a similar line, Khan et al. (2017) Chinese manufacturing firms, researchers collected the data of 415 firms from a different region of China, while they used regression analysis and factor analysis. The results show that green distribution and enterprise performance are negatively correlated, which means, due to the adoption of green practices in distribution, firms may lose their financial benefits. On the other hand, green distribution has a positive and significant impact on the competitive advantage of the firms, and due to the adoption of green practices in distribution, companies quickly can build long-term competitive advantage.

The green distribution contains efficiency of the distribution practices and exchange of information regarding sponsoring green consumerism and environmental management. Green recycling includes reuse, refilling, re-manufacturing products, and their packaging and packing. Liu et al. (2012) researched Chinese firms to recognize the importance of green distribution and its relationship with firms' performance and competitive advantage. The results confirmed that green distribution has a significant and positive association with enterprise performance and competitive advantage. In simple words, green distribution helps firms to build a competitive advantage in the market in terms of better image and reputation. On the other hand, green distribution has a nonsignificant effect on enterprise performance through competitive advantage.

Green distribution creates significant enhancement in the end-to-end financial performance of the firms. In the study, the association between green practices in SC and their effect on financial performance. The green manufacturing process may mitigate resource wastage and play a significant role in energy-saving, efficiently improving the supply chain financial and environmental performance with a positive corporate image. Khan et al. (2017) researched the green distribution and their effect on the firms' performance and competitive edge. Researchers collected the data from different manufacturing firms and applied structural equation modeling to test the hypothesis. The results show that green distribution has a significant and positive role in building competitive advantage for the firms in terms of image building and positive reputation, creates sympathy for firms in the customer's mind, and increases the consumers' market. Many types of research are conducted to analyze the relationship between green distribution practices and competitive advantage, and most researchers found that green distribution is not only correlated with competitive advantage but also has a significant and positive effect on the financial and environmental performance of the firms.

5.1.14 GD and ENP

The findings show that green distribution has a significant positive impact on environmental performance on a 1% confidence level. Furthermore, a 1% improvement in green distribution will increase a 0.17% increase in environmental performance. Green distribution has a critical role in the adoption of green supply chain management, and green distribution has a direct effect on the enhancement of environmental performance. Green distribution is an essential factor in the enhanced firm image in the customer mind, increased customer satisfaction, improved firm image and reputation, support of government and regulatory authorities, moral support of NGOs (nongovernment organizations) and the international community, attractive export opportunities, and more significant financial and environmental advantage. In simple words, green distribution plays a vital role in the implementation of green supply chain management. Similarly, Khan et al. (2017) researched the green distribution and their effect on enterprise environmental performance. The results show that green distribution has a positive and statistically significant impact on firms' environmental performance, while green distribution also helps firms to build a competitive edge for the long run.

Green purchasing and green distribution build a significant impact on the overall environmental performance of the firms (Zhou, Pan, Chen, & Li, 2015). There is no doubt that green distribution protects environmental sustainability, and it also reduces to the overall system cost in terms of reduction of waste, fewer steps in the distribution, shortest possible route to arrive at destinations, and usage of renewable energy sources. Liu et al. (2012) researched the central and south regions of China; researchers collected the data from different firms and applied structural equation modeling and factor analysis for testing their research's hypothesis. The findings show that green distribution has a positive effect on the firms' environmental performance and also support to build a competitive edge in the market. On the other hand, green purchasing is not directly effect to the firms' positive image. Because purchasing role is not in front of customers' eyes, distribution networks of firms are in front of customers' eyes, and consumers have their own opinions and remarks, and also green distribution build sympathy in customers' heart.

Carbon emissions, climate change, global warming, and global temperature may be controlled by the adoption of green practices in the distribution networks. Because of the substantial movement of transportation required in the distribution center and warehouse in terms of picking, storing, sorting packaging, asserting, break-bulking, cross-docking, and consolidation of products. Countries' image also increased by the adoption of environmental practices in the business and supply chain operations, while foreign direct investment inflows and per capita income are also correlated with green practices. Simpson (2012) the pressures of customers insist on firms to adopt green practices including green distribution and green environmental performance. Undeniably, the green distribution network is expensive as compared to the polluted distribution networks, due to expensive (fixed cost) green technology and employees of training. The research results show that green distribution is expensive when the starting cost is higher, but after a couple of years, it has minimized, and green distribution is more profitable than polluted distribution networks due to the elimination of waste, competitive edge built by firms based on customers' sympathy generated, positive reputation, and image of the firms. On the other hand, the tourism industry also booms due to the implementation of environmental-friendly practices and strict environmental policies implemented by the federal and provincial governments. In the perspective of Pakistan, during the last 4 years, the provincial government has done plantation of more than one billion trees in the province, which increased environmental sustainability and also attracted foreign direct investment inflows and foreign tourists. Climate change and global warming can only be controlled by practical action taken by regulatory authorities and firms as corporate social responsibility.

5.1.15 GT and CA

Green transportation and competitive advantage have a significant and positive relationship on a 10% level of confidence, while a 1% increase and improvement in green transportation will bring a 0.14% improvement in competitive advantage. In other words, green transportation plays a significant role in building a competitive edge in supply chain management. Besides, green transportation is protecting environmental sustainability and enhancing the firm's positive image and reputation in the market. In the adoption of green supply chain management, green transportation received minimal importance, but in the real case, green transportation has a significant contribution to the reduction of carbon emission, greenhouse gases, reduce global warming, control climate change and world temperature issues. Green transportation helps and provides an opportunity to create a competitive edge through adoption of best and green practices, while customers are more willing to buy firms' products if they know and physically assess firm transportation operations. Because transportation operation is straightforward to judge by customers and the firm's vehicle itself creates positive image building and reputation. Khan et al. (2017) researched green supply management practices' impact on corporation firms' performance. Researchers collected the sample data from different 415 manufacturing firms, including FMCG, leather manufacturing, pharmaceutical, food manufacturers, slow-moving consumer goods, etc. and applied exploratory factor analysis with multiple linear regression to test the research's hypothesis. Furthermore, researchers variable of green supply chain practices covering different sub-variables including, green transportation, green logistics, cooperation with customers green manufacturing, ecological design, and green information systems. The results show that green transportation has a significant effect on the firms' financial and environmental performance, while due to the adoption of green transportation practices, firms improved their image and reputation in their market and also enhanced the market share and market size. Furthermore, researchers also suggested that senior management should be well-committed towards the green supply chain project, which is a pivotal point to make the green project successful.

The nongreen transportation has a significant role in environmental degradation, and due to environmental pollution, a number of problems are occurring including health degeneration, global warming issues, killing of fauna and flora, and mainly climate change. These all are the key issues, which are very hot in the last couple of decades, and as per the international community pressures and United Nations' pressure, different government and legal authorities around the world are active and try to reduce environmental degradation and also promote renewable and clean energy systems to protect environmental sustainability. Sheu and Chen (2014) researched to investigate the incentive rivers used in many configurations in green transportation and also their effect on environmental sustainability. The researchers collected the data and used a different statistical test to examine the research hypothesis. The results show that green transportation management is not good only for environmental sustainability, but it has also enhanced profitability, while green transportation is the lowest cost as compared to nongreen transportation modes and practices. Researchers also emphasize that government and regulatory authorities should come on the front to pay their attention to this challenge and enforce the environmental policies for long-term environmental sustainability and reduce global warming/climate change. On the other hand, senior management should think over it not only in the context of profitability concerns but also from the environment perspective. Green transportation management enhanced the firms' profitability in terms of reduction of waste; renewable energy sources are inexpensive than nonrenewable energy.

Green transportation is a significant critical practice in the project of green supply chain management. Green transportation's objective is to increase environmental sustainability and control carbon, GHG emissions, and global warming. There are many factors includes in green transportation but mainly are; reduction of fossil fuel consumption, shifting vehicles on renewable/green energy/electrical/solar energy, proper maintenance of vehicles on periodically basis, providing suitable green training to the transport-related employees, use suitable and short routes to arrive destination/customers, consolidated delivery to customers, and buy fuel-efficient and advanced technology vehicles, which can not only reduce cost from the transportation system but also have significant improvement in environmental performance and grab competitive edge in terms of image and reputation building.

5.1.16 GT and ENP

The findings of our research confirmed that green transportation and environmental performance have a strong, statistically, and positive relationship on a 1% level of confidence. The results also revealed that a 1% increase in green transportation would enhance the practices in environmental performance. Undeniably, green transportation reduces the harmful effect of pollution, including carbon emissions and greenhouse gas emission on the environment, and also increases environmental sustainability. Many researchers prove that carbon emissions mainly emit from the transportation operation and more extended emission emits during global transportation due to long distances/lead-times. The fundamental objective of environmental-friendly transportation is to reduce the harmful effect of the supply chain and business operations on environmental sustainability while enhancing customer satisfaction and environmental performance with building a competitive advantage. The adoption of advanced green transportation dealing, which has less relation to cost savings and efficiency. Advanced green transportation has an essential role in the reduction of carbon emissions, climate change, and global warming. In a similar line Green et al. (2012) conducted research to identify the impact of green supply chain management practices on the performance of firms. The research paper empirically theorizes and assesses a comprehensive green supply chain management practices and model. The data is collected from 159 manufacturing firms in the USA, while the researchers used structural equation modeling to test the hypothesis. Results show that green supply chain practices, including green transportation, have a significant effect on the firms' performance and also have a significant role in the reduction of pollution. On the other hand, researchers also recognize that green transportation improved to the overall firms' financial position with positive image building in the market.

Khan et al. (2017) conducted the study to explore the effect of green supply chain practices on firms' competitive advantage and firms' performance from the perspective of China. Researchers collected the sample data of different firms in the central and south regions of China. The green supply chain management practices were measured by different green predictors, including green manufacturing, green transportation, green purchasing, etc. After using necessary statistical analysis for testing hypothesis, researchers identified that green supply chain management practices, including green transportation, have a significant and direct effect on the firms' competitive advantage building and environmental performance of the firms.

The implementation of an environmental-friendly system to enhance the environmental performance of the company as well as the intangible output such as customer loyalty and firms' image. Besides, green practices in the transportation sector not only improved to the firms' image in front of customers but also used to attract new customer markets to enhance market size. Furthermore, many European countries are discouraged to import polluted products through polluted transportation modes, which enhanced the green transportation sector. Nongreen products face nontariff barriers during the exporting process to the European countries. The practices of green transportation significantly improved to the firms' financial performance with environmental performance. Firms can control their business operations' harmful effect on the environmental sustainability through the adoption of green transportation practices, which are based on aerodynamic vehicles, and more fuel-powered vehicles are all adopted in terms to control their emissions and create a positive image, reputation building and also grab maximum opportunities in the international arena. The environmental performance mainly relates to the ability of production firms to minimize carbon emissions and solid wastes and the ability to reduce consumption of toxic chemicals and hazardous materials. On the other hand, green transportation cover and based on the reduction of fossil fuel consumption, reduction of energy and waste from the systems, usage of modern and environmental-efficient transportation vehicles, using green and renewable fuels such as low sulfur content and biofuels, providing proper green environmental training to the relevant employees, etc.

5.1.17 ED and ECP

The results findings show that ecological design and economic performance have a strong positive association on the 1% confidence level. 1% improvement in ecological design will bring 0.18% of the build competitive advantage for the firms. Undeniably, the ecological design has a positive association with a competitive edge and also helps in building image and reputation for the firms. On the other hand, the ecological design also reduces the cost of the end-to-end supply chain management in terms of different green practices and techniques including recycling, re-manufacturing, reuse, and refilling. The ecological design of products directly reduces the cost of the products in terms of recycling, reuse, re-manufacturing, refurbishment, and recovering materials and components from the product after the product life cycle (PLC). The first-mover benefits are where the company introduces green product design and enhances advantages covering getting royalties for licensing of green technology or products, unique production capabilities, and also create proprietary information which will lead to a sustainable competitive edge.

Khan and Dong (2017a) conducted the research to identify the green practices, including ecological design impact on firms' financial and environmental performance. Product stewardship incorporates in the context of external members into the ecological design development through the shelve life cycle; environmental products goods may measure the impact of the good on the environment. Firms introduce ecological design in the product, may improve financial profitability and environmental performance, and get royalties for licensing green technology, which will significantly lead towards green competitive advantage. In a similar line, Grønhaug and Kaufmann (1988) explain that ecological design has a positive effect on firms' competitive advantage. Many firms are using ecological practices into their product development stages in order to build competitive advantage and enhanced the positive image of the firms with financial and environmental performance (Govindan, Khodaverdi, & Vafadarnikjoo, 2015).

Companies introduce ecological design innovations to achieve advantages, including receiving royalties for creating proprietary information, which could provide a green competitive advantage. Product development and design processes may assess the environmental effects of goods till its last cycle of product stage. On the other hand, Zailani, Eltayeb, Hsu, and Tan (2012) claimed that the adoption of green practices improves enterprise financial performance, reputation, and image. In a similar line, ecological design not only increase firms' economic and environmental performance but also reduce harmful effects of manufacturing operations and minimize utilization of resource in terms of reuse, re-manufacturing, refilling, and recycling techniques (Mitra & Datta, 2014).

Companies emphasize ecological design may reduce the usage of toxic and hazardous materials in products and also may minimize resource consumption in the production of products. Besides, environmental-friendly product design is also facilitating recycling, re-manufacturing, reusing, and recovering components and materials from products after product life cycle. The green design of products has mainly significant effect on firm performance, and similarly, customers are more willing to purchase green products to attain the benefits of environmental protection and cost. The findings indicate that ecological design/green product design has a positive relationship with firm performance and creates a positive image and competitive edge, and the implementation of green design/environmental-friendly practices increased the company share market. Further, eco-friendly and green products reduce the cost of products and meaningfully improve the value of the product. Accordingly, we build the following hypothesis to measure product ED (ecological design) effect on firm ECP (economic performance). Khan et al. (2017) researched green practices of the supply chain, including the ecological design of products and their effect on the environmental performance of the firms and competitive edge. Researchers used the sample size of 415 firms' data to test their hypothesis, while the results revealed that ecological design has a definite strong association with a competitive advantage, and ecological design helps firms' to improve their image and reputation. On the other hand, researchers suggest that senior management should adopt green practices, but management staff should be well-committed towards green supply chain projects, because most times senior management is not showing their compelling interest in the project, which demotivates the employees and junior workers and staff, which results in a waste of money and time.

Green product design has a significant impact on a firm's survival and plays a vital and essential role of weapon to enhance environmental sustainability and create a sustainable competitive edge. Many firms are using green practices into their product innovation in order to create a competitive advantage and also differentiate themselves from other companies/competitors. Undeniably, ecological design practices are mainly dependent on supplier and customer management and senior leadership commitment towards green supply chain management practices.

5.1.18 ED and CA

The findings show that ecological design has a significant impact on building a competitive edge for firms, while a 1% improvement in ecological design will bring a 0.23% increase and improvement in a competitive edge. In simple, ecological design is regarding environmental protection and also increased to the profitability for firms in terms of different techniques, including recycling, re-manufacturing, refilling, reuse, etc. On the other hand, ecological design is necessary for today's world due to pollution issues and global warming challenges. Many firms are implemented and also adopting ecological design in their product development stages. However, they still required further improvement and encouragement from the governmental and regulatory authority sides to promote eco-friendly design products.

Firms may build a competitive edge from the adoption of ecological design in their products. The first-mover edge is where the company introduces eco-friendly products innovation in items design and enhance advantages, including getting royalties for licensing developing unique production capabilities. Ecological product design creates a competitive edge for the firms in terms of positive brand image and improved firms' reputation. Ecological product design needs that producers design to the product, which may reduce consumption of energy, materials, and components that facilitate the firms in terms of recovery of materials and parts, reuse, recycle, and re-manufacturing or refurbishment.

In a similar line, Grønhaug and Kaufmann (1988) explain the green product design impact on the company and have a significant effect on environmental sustainability and also increased the competitive edge. Khan et al. (2017) researched green practices and ecological design effects on manufacturing firms' competitive building. Researchers collected a sample of 415 firms from the different areas of China to test their hypothesis. Green practices and ecological design are measured, including green distribution, green ecological design, green manufacturing, and green purchasing. The results confirmed that green purchasing has no such significant effect on the firms' performance and competitive edge building, while remaining exogenous variables including green manufacturing, green distribution, and green ecological design have a significant effect on the firms' performance and building competitive advantage. Finally, researchers also realized that ecological design has a direct and superior effect on the competitive edge building, due to ecological design firms, and also may reduce end-to-end supply chain cost through using recycling, re-manufacturing, refilling, and reuse techniques, which is easy for green products as compared to nongreen products.

Khan and Dong (2017a) conducted the research on the green supply chain practices impact on the manufacturing firms; they found that ecological design not only reduces harmful effect on the environmental sustainability but also enhanced firms' performance in terms of recycling, re-manufacturing, and reuse approaches; the further green design has a positive and significant impact on the organizational performance. Organizations focus on environmental-friendly design may reduce the usage of hazardous materials and chemicals from the products and also minimize resource consumption in production goods.

Furthermore, ecological design products enhance enterprise performance, and customers are also satisfied with firms, which create a competitive edge for the firms in terms of positive brand image and reputation in the domestic market and international markets. Besides, eco-friendly design products create uncountable export opportunities for firms. Many motivating factors create a direct and significant effect on the adoption of green supply chain practiced and found that ecological design has a positive influence on the firms' performance. The principal objective of ecological design is to minimize products' harmful effects on the environment without reduction of firms' internal profitability.

5.1.19 ED and ENP

The findings revealed that ecological design and environmental performance have a strong positive relationship on the 1% confidence level. In simple words, the ecological design will increase, and such an environmental performance will be improved further, and a 1% improvement in ecological design will lead to a 0.29% improvement in environmental performance. It means that whenever ecological will be increased from 1 percent, so the environmental performance of the firms will also increase from 29%. Undeniably, the ecological design of products has many fruitful benefits for the organization in terms of cost reduction, process efficient, fewer steps of manufacturing, and total quality management. Green product design is a vital part of the concept of green supply chain management; the main ecological design reduces environmental pollution in terms of carbon dioxide, greenhouse gas emission, global warming, climate change, and world temperature. The manufacturing firms create several problems for the environmental sustainability and increased air pollution and water pollution, while ecological design talks about environmental sustainability and the greener environment with improved firms' performance and creates a competitive edge in terms of positive image building and reputation.

The adoption of green supply chain management practices is expected to result in enhanced firms' environmental performance as measured by effluent waste, consumption of toxic chemicals and materials, solid waste, and reduction in air emissions. On the other hand, some environmental efforts will finally convert into a competitive edge, environmental performance, and financial performance of the firms.

The environmental performance of the firms relates to the ability of production plants to minimize air emissions, solid waste, effluent waste, and the ability to reduce consumption of toxic materials and chemicals. Zhu and Sarkis (2004) researched the Chinese market to identify the relationship between green supply chain practices and firms' performance. They found a robust positive association among the implementation of ecological design supply chain and improvement in the environmental and financial performance of the firms. Geffen and Rothenberg's

(2000) manufacturing setting, healthy relationships, and close collaboration with the suppliers result in enhanced environmental performance. Preuss (2001) describes the possibility of a "green multiplier effect" that results from the extension of green purchasing practices from immediate suppliers to second- and third-tier suppliers' suppliers.

Green supply chain management practices are built specifically to enhance the environmental performance of firms. Practices, such as cooperation with customers, ecological design, green purchasing, and investment recovery, are designed to improve the environmental performance of the firms.

The goal of ecological design products is the reduction of a product's environmental impact without creating any harmful effect on environmental sustainability and business opportunities, such as costs and functionality. Environmental-friendly product design and an emphasis on investment recovery will positively and directly influence the environmental performance as the push of the designers will be on decreasing the environmental impact of the design.

5.1.20 ENP and ECP

The findings show that the environmental performance of the firms has a statistically positive and significant effect on the firms' economic performance on the 1% level of confidence. Further, a 1% improvement in the environmental performance of the firms will lead to a 0.31% improvement in firms' economic performance. In simple words, environmental performance not only protects environmental sustainability, but a higher level of environmental performance also enhances firms' economic performance in the long run. The adoption of green supply chain management practices directly increased environmental performance and sustainability but also enhance financial performance, and indirectly green supply chain focuses on waste elimination related to environmental sustainability. Preuss (2001) describes the possibility of a "green multiplier effect" that results from the extension of green ecological design and purchasing practices from immediate suppliers to second and third-tier suppliers' suppliers. Based on anecdotal evidence, argue that green supply policies are likely to result in enhanced environmental performance. The practices of green supply chain management are explicitly developed to enhance the environmental performance of manufacturing companies. Green practices, such as ecological design, green manufacturing, green purchasing, investment recovery, cooperation with customers, and green distribution, are designed to improve environmental performance.

Many studies have been conducted to identify drivers that affect the adoption of green practices of supply chains; many studies found that green design positively influences the performance of supply chains. The principal objective of ecological design is the reduction of a product's environmental impact without negative trade-off with other design criteria, such as costs and functionality. Ecological product design and an emphasis on the recovery of investment will directly and positively influence environmental performance.

Generally, economic performance refers to the production plant's ability to minimize costs related to consumption of energy, purchased materials and components, waste discharge, waste treatment, and heavy penalties and fine for environmental accidents. Khan et al. (2017) conducted research to identify the impact of green supply chain management practices on environmental performance and the effect of environmental performance on firms' financial performance. Researchers collected a sample of 415 manufacturing firms in Pakistan and used linear regression to test the hypothesis. Researchers measured green supply chain practices by green distribution, green manufacturing, ecological design, and green information systems. The results revealed that green supply chain practices are significantly and positively enhanced to the environmental performance, while environmental performance also significantly and positively improved to the firms' economic performance in the long run. In simple words, green supply chain practices are not only protecting environmental sustainability but also enhanced firms' financial performance in the long term. Besides, researchers also identified that firms also build competitive advantage through the adoption of green practices in their supply chain and business operations. On the other hand, researchers also suggested that, during the implementation of green supply chain projects, senior management should be well-committed and show their interest in green projects. Because without the seriousness of top management, green projects cannot achieve their essence.

The cost-saving nature of environmental performance needs to achieve and enhance financial performance and build a competitive edge with positive image building and firms reputation in the international market, while financial performance also enhances operational performance and reflects firms' ability to satisfy customers demand of green products and services. The marketing implications of environmental performance lead to significant enhancement in economic performance.

5.1.21 ENP and ORP

The results show that environmental performance has a positive and significant relationship with organizational performance on the 1% confidence level while on a 1% improvement in environmental performance will bring 0.77% improvement in the overall organizational performance. In simple words, environmental performance is not only protection to environmental sustainability but also enhanced the overall supply chain and organizational performance and increase to the market share and size. On the other hand, the environmental performance also allows firms to build a competitive edge in the market, while many European countries have banned non-green products in their country, so ultimately, nongreen products have become a nontariff barrier for the polluted manufacturers.

The cost-saving nature of the environmental performance should achieve to enhance both environmental performance and economic performance, while financial performance should yield an increase in the operational efficiency of the supply chain management and the firms. Environmental performance creates cost-saving and reflects a firms' ability to fulfill changing customers' requirements for environmental sustainable services and products. The marketing implications of environmental, operational, and economic performance should lead to enhancement in the end-to-end marketing and financial performance of the firm.

Undeniably, environmentally friendly practices increased the positive effect on the environmental performance of the firms, and it also enhanced organizational performance. The cost-saving nature of environmental performance should attain to increase both economic and operational performance with overall organization performance. Green practices implementation in the supply chain generate cost-saving and echo a firm's ability to fulfill the demand of customers for green products and services. The marketing and cost implication of operational and environmental performance may lead to a significant contribution to firms' overall performance. Green et al. (2012) conducted research on the manufacturing firms of the USA, while researchers collected the data of 159 different manufacturing sector firms and used structural equation modeling to test their research's hypothesis.

The results revealed that green supply chain management practices improved to the firms' performance, but also senior management and green information systems are the backbone of the green projects. Due to the lack of interest of senior managers, green projects can be converted only into a drain on firms' profit. On the other hand, environmental performance increased to the overall organizational performance and also created a competitive edge for the firms in the long run. Khan and Dong (2017a) collected the data from Pakistani manufacturing firms and used linear regression and some other statistical analysis to test their research hypotheses. The researchers used ecological product design, green manufacturing, green purchasing, green distribution, and cooperation with customers as green practices variables. The results revealed that except green purchasing, the rest of green practices, including green manufacturing, eco-friendly product design, green distribution, cooperation with customers, all have a significant and positive effect on the firms' performance. Finally, researchers also disclosed that green project success mainly depends on the senior management commitment towards green supply chain projects and green information systems to monitor and evaluation to the green project.

Once green practices have become strategic attention and (information systems) have been modified to monitor and measure the environmental efforts of firms, manufacturers can begin to adopt green practices with some extent of confidence that the practices will yield increased environmental performance but also enhanced overall company performance. The execution of green supply chain management practices enhances the overall companies' capabilities to sustain the environment and to support the overall firms' performance.

5.1.22 CA and ORP

The finding shows that competitive edge and organizational performance have a strong positive association on the 1% confidence level, while a 1% increase in the competitive advantage will lead 11.3% improvement in the overall organizational performance. In short words, competitive advantage is built by green practices and green techniques, which also enhance organizational performance. Green supply chain management practices create a competitive advantage in the domestic and international markets, while competitive edge creates a significant and positive effect on the overall organizational performance. The results show that competitive advantage build by green practices including, green manufacturing, green design/ ecological product design, green transportation, green warehousing, and green material management, which create a significant and direct impact on the building competitive advantage for the firms and ultimately competitive advantage enhanced firms' overall performance.

Kushwaha GS and Sharma NK (2016) claimed that implementation of eco-friendly techniques and practices in the supply chain management would significantly lead firms' performance, and execution of environmental-friendly practices not only enhance the environment but also build a competitive edge. Many researchers researched to show that green supply chain management practices significantly build a competitive edge for the firms, while competitive edge and green practices both enhanced the firms' performance. Hart (1995) recommended that incorporating green ideology into a company's strategic planning will increase the firm's ability to overcome environmental challenges, reduce overall system cost and uncertainties at the confluence of green issues and companies operations, but also support and lift firms to build competitive advantage, increasing their economic, operational, and environmental performance.

An enterprise's green supply chain management can be regarded as a capability for acquiring a green competitive edge and may have a positive effect on the company's and even SC's (supply chain) performance and competitiveness although many researchers have demonstrated that green supply chain management is significantly and positively correlated with a competitive advantage, while green technology and practices also directly and indirectly correlated with the economic and environmental performance of the firms. Liu et al. (2012) this research paper conducted to identify the effect of green supply chain ideology including, green distribution, green manufacturing, green purchasing, green logistics, and ecological design products on competitive advantage and organizational performance, while researchers collected the sample size from the central and south region of China. After using exploratory factor analysis and structural equation modeling, the results confirmed that green practices have a strong positive association with firms' performance, on the other hand, green practices also help to build competitive advantage for the firms. Finally, researchers also realized that competitive advantage has a positive effect on organizational performance. It means a higher level of the competitive edge will lead to a higher level of organizational performance.

Green manufacturing, green design of products, green distribution, and green logistics are the key pillars of green supply chain ideology, while green information system plays a role as mediator and/or interchangeable technology, and internal environmental management is the backbone of the green project in the supply chain operations, which not only help to show the success of the projects but also show how to succeed and what should be next and how management should decide to improve the effectiveness and efficiency of this project. Many researchers have confirmed that green supply chain management approaches including green distribution, green manufacturing, green logistics, green ecological design, green packaging, and packing have a significant and positive effect on building competitive advantage along with the enhanced economic performance of the firms. On the other hand, green purchasing sometime in some industry has significant positive association with competitive edge while in some industries have as such no correlation with competitive advantage, e.g., a research conducted by Khan and Dong (2017a) shows that green purchasing is negatively correlated with firms' organizational performance, due to as such no incentives offered by government and regulatory authorities to support firms' green initiatives.

Organizational performance refers to the marketing performance and financial performance of the firms as compared to the industry average firms' performance. The findings confirmed that green practices enhance environmental performance and build a competitive advantage in the market as compared to their competitors. Finally, it enhanced the overall organizational performance/firm performance.

5.1.23 ECP and ORP

The results indicate that the economic performance of the firms has a positive and robust relationship with organizational performance on the 1% confidence level. Besides, a 1% improvement in economic performance will achieve an 11.2% improvement in organizational performance. It means that economic performance not only improves the firms' wealth but also increases competitive environment, environmental performance, and operational performance, which leads to overall organizational performance. The adoption of green supply chain management practices mainly including green purchasing, ecological design, cooperation with customers and suppliers, green manufacturing, green distribution, and transportation leads to higher economic performance and competitive edge, while the economic performance of the firms enhanced overall organizational performance. According to Liu et al. (2012), the research results show that the economic performance of the firms has a strong positive association with the overall supply chain and organizational performance of the firms in the context of China. Besides, firms also build a competitive edge through the adoption of green practices in the supply chain, which significantly enhances the firm's positive image and reputation in the market. Green manufacturing reduces the waste from the systems and also eliminates unnecessary steps from the manufacturing systems and increases to the efficiency of the manufacturing along with an improved enterprise image and reputation (Porter and Linde 1995). Firms in the USA and European regions have incorporated and integrated total quality management philosophy and lean manufacturing practices with ecological product design and manufacturing to attain a more superior economic, which also enhances the overall supply chain and organizational performance.

Environmental-friendly manufacturing plays a significant role in adopting green supply chain techniques to control the adverse effects of manufacturing operations with little waste in overall manufacturing systems and also helps firms to improve efficiency in the processes. The environmental-friendly purchasing includes recycling, reuse, and re-manufacturing in the supply chain, which also enhances the environmental sustainability and control of the firm's overall cost and creates a positive image and reputation in the market. Undeniably, the economic performance of the firms also enhances the market share, investment returns, and return on assets, which are the parts of organizational performance. Khan and Dong (2017a), the first-mover advantage, is where the firm introduces ecological product design and enhances advantages, including adopting unique production capabilities, and getting royalties for licensing green innovation, and also build proprietary information which significantly achieves to the environmental performance and competitive advantage. Similarly, Grønhaug and Kaufmann (1988) explain the ecology design benefits and their effect on the firms' environmental performance, operational performance, and, finally, organizational performance in the long run.

The adoption of green techniques and practices, top management support, is a crucial driver of the successful implementation of green technologies. Once firms adopted green practices and techniques as a strategic imperative, then they may continue to build green information technology capabilities. Firms can face problems if they compare economic performance with environmental performance and organizational performance due to poor data management. The research finding shows that investment in green technologies is significantly connected with firms' performance. However, without proper and in-depth understanding, technology may be converted into a drain on firms' profitability.

Green logistics, green manufacturing, green distribution, green eco-friendly product design, and green purchasing create significant improvement in overall firms' environmental and financial performance. Studying the association between green practices and techniques in supply chain and companies' performance has, directly and indirectly, a significant role in companies' performance. Porter and Linde (1995) confirmed that environmental-friendly manufacturing process might minimize the harmful effect on environmental sustainability and also reduce the waste from the overall systems, have fewer manufacturing steps, and significantly enhance companies' overall performance, image, and reputation in the international arena. Green distribution helps companies to accomplish environmental performance financial performance and organizational performance.

The green technology investment enhances environmental performance but also increased financial performance. Also, the economic performance of the firms leads to enhanced organizational performance. Several companies implementing green practices to improve their environmental and economic performance including

Engro Corporation, General Motors, and Procter and Gamble also invest $3 billion for implementing advanced technology and green practices in their business to improve performance (Khan & Dong Qianli, 2017), because they know economic benefits will directly and/or indirectly improve overall company performance in the long run.

Based on the above discussion, the researcher provides concluding remarks on this study with innovation points and suggestions. Mainly researcher results and recommendations are specifically for the country of Pakistan. Because the researcher collected the sample data from different areas of Pakistan and according to the data all the analyses and results have been discussed, most suggestions and recommendations are not generalized.

5.2 Qualitative Part of the Research

In this part, we will discuss the qualitative part of this research through the findings based on unstructured interviews with senior management of manufacturing firms of Pakistan. We also distributed results of the unstructured interview into two main categories, namely, barriers and drivers. Further, we will discuss each category in the context of competitive advantage, economic performance, environmental performance, and organizational performance. The following discussion has been mentioned with crucial results of research and also supported by other researchers' works.

5.2.1 GSCM Barriers in Manufacturing Firms of Pakistan

From the perspective of Pakistan, many firms are adopting green supply chain management practices in their different activities and operations. Undeniably, during interviews with senior management of manufacturing firms of Pakistan, we have identified several barriers, which not only create hurdles for firms to acquire green knowledge but also creating issues during the adoption of green practices. The barriers of GSCM in the manufacturing firms of Pakistan mainly have been divided into the following categories for better understanding.

5.2.1.1 Economic Barriers

There are many economic barriers faced by manufacturing firms of Pakistan during the adoption of green supply chain management practices. We will discuss some vital economic barriers in the following paragraph for the development of understanding.

For the most part, enterprises struggled with financial constraints to start green projects and adopt ecological practices in their companies' supply chain management operations. Undeniably, the administration of the organic supply chain required huge fixed costs for the initial configuration of green operations without problems, as companies need to buy advanced technologies to reduce emissions and waste. Besides, activities related to recycling and regeneration also require flexible product design, which is possible only with the use of advanced technology in the activities of enterprises. Some companies are afraid of using ecological practices in the supply chain due to the enormous capital expenditure required in the initial phase of green projects. On the other hand, several companies have confirmed that after the adoption of environmental supply chain practices in the operations of their enterprises, the variable cost of the end-to-end supply chain is significantly lower than traditional operations in the supply chain. The traditional supply chain is based mainly on contaminated practices, which not only increase the consumption of materials and chemicals but also reduce the environmental efficiency of companies with negative image and reputation of companies.

High-tech systems support cooperation processes in the supply chain and improve supply chain efficiency. An efficient IT and technology system is very much needed to support GSCM at various stages of the product life cycle. It can be advantageous in product development programs that include environmental design, recovery, and reuse. To track and track product returns, you need adequate information systems related to your previous sales (Ravi & Shankar, 2005). Information support is needed to develop links to achieve effective GSCM in manufacturing companies in Pakistan. It is necessary to manage the flow of information related to the flow of materials and other resources back and forth to effectively manage green SC (AlKhidir & Zailani, 2009). Besides, enabling IT reduces paper consumption, which supports the GSCM philosophy. Then, the lack of IT implementation is an essential obstacle to achieving an efficient GSCM.

Technology is a kind of knowledge. An organization with extensive experience in the application and implementation of related technologies will have a higher capacity for technological innovation. The organization will have a more exceptional innovative ability when knowledge can be more easily shared within the organization. Technological progress can be achieved through greater portability. It is easy to share technology transfer or share technology knowledge (Cooper, 1994). Innovation and technology contain innovations in corporate culture, supporting new ideas and processes and solutions by all employees of the company. The organization's resistance to the adoption of technological progress is resistant to change. The organizational barrier means difficulties in achieving a fundamental change in the organization. This is especially important when changes occur in the main features of the organization, such as organizational goals, forms of power, underlying technology, operational strategy, and market strategy. Therefore, resilience to adopting technological advances is an essential obstacle to the implementation of GSCM in the automotive industry.

Historically, the cost was used as the primary measure of performance. In general, high cost is enormous pressure on GSCM compared to conventional SCM. Initial investment requirement in the field of ecological methodologies, such as ecological project, ecological production, eco-labeling of packaging, etc. It is too high. Participation in environmental management includes two types of costs, direct costs and transaction costs. Both types of costs are likely to be a significant obstacle to the implementation of GSCM. Enabling IT, adopting technological progress, employing good quality employees, motivating, and training employees towards GSCM will require a substantial initial investment. Therefore, the cost implication is a significant barrier between barriers hindering the implementation of effective GSCM in manufacturing companies in Pakistan.

They are strengthening relationships with suppliers that result in lower inventory levels, costs, and higher precision. The participation of suppliers in the design and technology process affects the overall results of the entire chain. Suppliers' supplier relationships are considered more important to gain a competitive advantage for the manufacturer. Large production companies usually have from 2000 to 3000 suppliers. Producers cannot produce organic products unless they cooperate with suppliers. Suppliers must comply with buyers' requirements to maintain business relationships. Therefore, we can say that the reluctance of suppliers to move to GSCM is an obstacle in the implementation of GSCM in manufacturing companies in Pakistan.

Government incentives and incentives can encourage innovation because the government praises and motivates companies. Time-consuming regulatory requirements, fees, or charges may discourage smaller companies. Tax structures that distort incentives may discourage the industry from implementing GSCM. Governmental institutions are seen as obstacles in the development of environmental management in the sense that the institutional process of implementing GSCM is underway, but minimal institutional support was provided for new ideas for the implementation of GSCM. The government's tendency to encourage old practices is an obstacle (AlKhidir & Zailani, 2009). Therefore, the lack of government financial support systems is a barrier to the implementation of effective GSCM in manufacturing companies in Pakistan.

The main barrier of GSCM observed in manufacturing companies in Pakistan is the lack of awareness of customers about the benefits of organic products. The clients' requirements are becoming the most critical type of external pressure. Client's awareness means that the customer demands ecological products; the company needs to change technology and organization for innovative organic products. However, in manufacturing companies in Pakistan, due to the unconsciousness of customers regarding the benefits of organic products, some manufacturers produce nonorganic products. In the USA, it is estimated that 75% of consumers say that their purchases have a reputation, and 80% are willing to pay more for environmentally friendly products. It can, therefore, be said that the clients' unawareness is the main obstacle to the implementation of GSCM in manufacturing companies in Pakistan.

Lack of knowledge in society means that customers are still not aware of organic products and their benefits to society. Walker, Di Sisto, and McBain (2008) found that there are many complications, such as fear of inadequate disclosure of environmental performance, lack of confidentiality and information, lack of strategic planning; insufficient industrial structure, lack of knowledge in the field of technology and management; lack of qualified workforce, lack of public participation, high investment risk, lack of appropriate production technology, lack of support for energy saving in China, lack of financial support and difficulties, insufficient data and information, group opposition created are the main obstacles for GSCM. Each industry has a different barrier to GSCM adaptation, which depends on the type and nature of the company (Walker et al.). No public pressure. Low social pressure is a barrier to the implementation of green manufacturing. Lack of pressure from social sectors such as local communities, media, NGOs, banks, insurance companies, or politicians may not provide the necessary pressure for companies to innovate ecologically. Massoud, Fayad, Kamleh, and El-Fadel (2010) discovered various barriers to the implementation of ISO 14001 and the environmental management system (EMS) in the Lebanese food industry. These barriers include demand for time, lack of knowledge, lack of priorities in management, cost of certification, lack of demand from customers, not required for export, lack of government support, lack of clear benefits, extra-legal requirements, and lack of government support.

Lack of knowledge about GSCM among stakeholders in the supply chain is considered one of the biggest obstacles to the implementation of GSCM. This means a lack of experience and a sense of "too complex" implementation. Green implementations include advanced technology and specific knowledge, and organizations without this knowledge require proper training. Ravi and Shankar (2005) also emphasized that training and education are vital requirements to succeed in implementing GSCM practices in any organization. These training and education involve costs in terms of time and money (Khiewnavawongsa & Schmidt, 2013).

If organizations want, they should provide understanding to people who will have to do it practically. Participants often mentioned culture as one of their most significant barriers. The mentality of Mozambique is not conducive to GSCM, and care for the environment is not rooted in their way of thinking. This is due to the amount of garbage on the streets. The following interview quotes indicate that culture is the main barrier in Mozambique's production organizations: "People prefer to hire an independent professional who will be less burdened, but has more environmentally harmful trucks." "Culture about the environment is so strong if the people responsible for its implementation have a different culture due to their education and origin.

All organizations incur costs, and most, if not all, are solely aimed at reducing them to increase profits. Consumers want to reduce costs in industries, forcing a limited budget for producers and reducing interest in green investments (Walker et al., 2008). Concerns about costs were a significant obstacle in five of the eight organizations surveyed because these organizations focused more on short-term benefits than on the long-term benefits of participating in GSCM. Organizations did not want to bear the initial costs necessary for greener practices. Although participants have clearly stated that they have implemented environmental practices based on cost reduction, most of these practices are those that require low initial investment. One participant explained that costs are a significant problem in the manufacturing industry in Mozambique and that they are always seeking to reduce it. Participation in GSCM is desirable, but no one wants to make the initial investment required. These observations can be observed based on quotes from two participants: "GSCM practices involve costs that act as a massive barrier because it requires monitoring, time and dedicated people and massive investments. 'Besides, it is a price-sensitive market, and adding value to price increases is not valuable.' Investors looking for Africa to get higher profits to see more short-term benefits."

Corruption refers to acts in which the power of a public office is used to obtain personal gain in a manner contrary to the rules of the game (Jain & Sharma, 2014). This leads to clean waste and improper allocation of resources (Aidt, 2009). Participants believe that Mozambique is a country trapped by corruption and says that this is a big obstacle. One of the eight participants highlighted the role of corruption as an inhibitor of GSCM practices. The participant mentioned that even if there are rules that allow organizations to respect the environment more, the obstacle can be eliminated simply by bribing officials. This not only discourages producers who invest and participate in greener practices but also motivates dishonest and unethical producers to gain an advantage over the competition. The literature on corruption is minimal, as a barrier to implementation in GSCM. It is an opportunity to conduct more research on the impact of corruption on the implementation of GSCM practices, especially considering that this is an African reality.

The government can act as a driving force and barrier. Almost all participants noticed that it is necessary to change the vision of the government in terms of environmental aspects. Seven participants mentioned that the government is the driving force behind the implementation of green practices in the supply chain, although it is still in the development phase of the child. Routroy (2009) argues that governments should participate in activities aimed at educating and promoting GSCM practices. Participants mentioned that the government has recently established meetings with producers to discuss greener practices. It is a factor that contributes to the awareness and significance of GSCM. The Government of Pakistan is seeking to enforce a limited number of environmental laws. "We have meetings, and the government says that please take into account, do things in this way and grant a period of completion, but it is not yet compulsory." This is in line with the literature review, as the researchers suggest that governments have a strong driving force in implementing GSCM practices. The current lack of government support means that the government does not provide policies that benefit the industry or provides unique benefits to organizations implementing GSCM practices (Luthra, Kumar, Kumar, & Haleem, 2011). Respondents often mentioned the lack of legislation and incentives to motivate the organization. Two participants mentioned that the government is slowly beginning to see the need and importance of green supply chains. However, this is the early days for the country, and much work still lies ahead.

Customers play a vital role in adopting and hindering the implementation of GSCM practices. Producing economic goods at the expense of the environment, which forces the industry not to accept GSCM practices. Lack of demand and public awareness are a severe problem when implementing GSCM. Lack of understanding and knowledge among suppliers is an obstacle to the implementation of GSCM practices. The supplier's failure/unwillingness to switch to GSCM (lack of willingness to exchange information), lack of understanding among stakeholders in the supply chain, etc. are barriers related to suppliers. Lack of knowledge about environmentally friendly goods and services in society is an obstacle to the adoption of ecological awareness-raising practices. Lack of knowledge in society means that customers are not yet aware of organic products and their benefits to society. The research found that various industries have various barriers to the implementation of ecological practices. Undoubtedly, various industrial sectors have different barriers to adopting GSCM practices. In the previous discussion, we listed the main barriers for manufacturing companies in Pakistan.

5.2.1.2 Environmental Barriers

There are many environmental barriers faced by manufacturing firms of Pakistan during the adoption of green supply chain management practices. We will discuss some critical environmental barriers in the following paragraph for the development of clear understanding.

Environmental barriers create obstacles when adopting green supply chain management practices in enterprises, while several companies overcome these barriers through their experience, experience of suppliers and distributors, senior green project consultants, etc. For the most part, these barriers become the main obstacle for a company that does not want to go to ecological operations. Due to the lack of information and knowledge, the lack of consumer awareness causes less pressure and support in the implementation of the organic supply chain due to lack of knowledge about customers, stakeholders, and suppliers, as well as due to the lack of commitment of senior management, due to the lack awareness that investors and shareholders cannot push for the organization of GSCM implementation. Ignorance of consumers towards an ecological environment. Knowledge: Lack of knowledge among people interested in the organization, Lack of knowledge of sellers, suppliers on GSCM, Suppliers do not know the destructive consequences of their product for the environment, Suppliers lack knowledge about the damaging consequences of your product in the environment. Legislation: therefore, no concerns and legislation regarding environmental protection, lack of professionalism, and legislation, there are no regulatory guidelines on environmental management by GSCM. SOPS: No SOPS human resources to develop GSCM. Competitors: Market competition for generic drugs is essential because of the traditional mentality regarding fear of failure. The lack of funds for global marketing, the cost of environmentally friendly products and packaging is very high. Resources: no resources to deal with global marketing, due to the lack of resources in industries for proper recycling, due to the lack of resources in the industries for proper recycling. Region: Lack of green architects, consultants, green developers, contractors in the region, in the region there are no bank loans available for industries in which you can implement environmental practices, pressure on timely delivery of the product in the region, because there is no market for recycled products in the region, due to the lack of internal audits in the organization.

Technology is a kind of knowledge. The organization will have a more exceptional innovative ability when knowledge can be more easily shared within the organization. An organizational barrier means difficulties in implementing a fundamental change in the organization. This is especially important when changes occur in the main features of the organization, such as organizational goals, forms of power, underlying technology, operational strategy, and market strategy. Therefore, the lack of acceptance of progress in new technologies is an essential obstacle to the implementation of GSCM in the small Pakistani industry. Legal requirements, fees, or time-consuming fees may discourage smaller companies. Tax structures that distort incentives may discourage the industry from implementing GSCM. Government institutions are considered to be obstacles to the development of environmental management in the sense that the institutional process of implementing GSCM is ongoing, but there is minimal institutional support for new ideas for the implementation of GSCM. Therefore, the lack of government support systems is a barrier to the implementation of effective GSCM in Pakistani manufacturing enterprises. In general, administrative support is a critical element in the adoption and implementation of innovation in the organization, especially in environmental systems. Organizational innovations may remain stagnant in the initial phase of the idea without dedicated champions. Support of senior management can influence the success of new system initiatives by promoting employee empowerment, facilitating employee participation by promoting cultural change, and greater involvement of employees in the organization, and introducing incentives and incentive systems to affect employees. Suppliers' pressure and will is a GSCM driver that determines that suppliers can help provide valuable ideas for environmental projects, but generally, do not act as a direct driving force. Although suppliers may not be drivers, integration and cooperation in supply chains can support more efficient management of environmental problems. A cooperative paradigm has been used to examine the practices of green supply chain management in production facilities. We can, therefore, say that the flexibility of switching to GSCM is a significant obstacle to the implementation of GSCM in manufacturing companies in Pakistan.

Training and education are the main requirements to succeed in implementing GSCM in any organization. Management can encourage employees to learn ecological information. Organizations can provide rewards for green employees. Employees can get help when they face ecological problems and can get help in learning ecological information (Hsu, 2008). Lack of GSCM training is one of the primary barriers to implementing GSCM in manufacturing companies in Pakistan. Previous studies have developed supply forecasts, but they were not designed explicitly for sustainable SCM. For example, 5-year forecasts for the offer show that scientists are trying to look to the future (Mudgal, Shankar, Talib, & Tilak, 2009). The studies brought together professionals and academics to predict the future of orders and deliveries within 20 years in the future. There are forecasts regarding the future of consumer demand for organic products. It reflects the integration of all departmental affairs related to the coordination of the implementation of barriers to green supply chain management in the small Pakistani industry. Previous studies have developed supply forecasts, but they were not designed explicitly for sustainable

SCM. For example, 5-year forecasts for the offer show that scientists are trying to look ahead. The studies brought together professionals and academics to predict the future of orders and deliveries within 20 years in the future. There are forecasts regarding the future of consumer demand for organic products. It reflects the integration of all departmental affairs related to the coordination of the implementation of barriers to green supply chain management in the small Pakistani industry.

Customers are becoming more and more educated and aware of environmental degradation and want the product they buy to be respectful of the environment. They may also wish that the final disposal and decomposition takes place without harming the environment. Customers in developed countries are more aware of this problem than Pakistani clients. That is why companies that export their products are more suited to adopting green practices in the supply chain, according to the requirements of their customers. Recent research indicates that the interface with clients provides valuable information that helps implement GSCM in the organization. Zhu et al. (2008) discuss the possibilities of GSCM in Chinese companies in different contexts and call for greater cooperation between companies and clients to adopt GSCM practices. In recent studies, Kumar, Chattopadhyaya, and Sharma (2012) focused on the emerging economy of developing countries and proposed a model that combines environmental policies with customer activities and sustainable development strategies. They found that customer management concerning eco-design, product recovery, and reverse logistics is crucial to achieving GSCM. Therefore, customer relationship management is an essential factor in facilitating the implementation of GSCM.

The government can act as a driving force and barrier. Almost all participants noticed that it is necessary to change the vision of the government in terms of environmental aspects. Seven participants mentioned that the government is the driving force behind the implementation of green practices in the supply chain, although it is still in the child's development phase. Routroy (2009) argues that governments should participate in activities aimed at educating and promoting GSCM practices. Participants mentioned that the government has recently established meetings with producers to discuss greener practices. It is a factor that contributes to the awareness and significance of GSCM. The Government of Pakistan is seeking to enforce a limited number of environmental laws.

"We have meetings, and the government says, please, take note, do things this way and give some time to end it." This is in line with the literature review because scientists suggest that governments are driving the United States in implementing GSCM practices. The current lack of government support means that the government does not provide industry-friendly policies or grant unique benefits to organizations that implement GSCM practices (Luthra et al., 2011). Respondents often mentioned the lack of regulations and incentives for motivating organizations. Two participants mentioned that the government is slowly beginning to see the need and importance of green practices in the supply chain. These are the first days in the country, and there is still a lot to do; six of the eight participants complained about the lack of incentives. These participants suggested that if incentives were established, the barrier would become a driving force. This lack of support is an excellent inhibitor, and several organizations are fighting for incentives to start a business. We fight for government incentives to ensure that our competitors and we are involved in initiatives related to green supply chains. We have an association that fights for tax breaks for companies that implement them. Research contradicts the findings of the literature review, as the government is considered to be one of the main factors influencing the implementation of GSCM practices. In the case of Pakistan, the lack of legislation and government support was listed as one of the biggest obstacles to the implementation of GSCM.

They are strengthening relationships with suppliers that result in lower inventory levels, costs, and higher precision. The participation of suppliers in the design and technology process affects the overall results of the entire chain (Mudgal et al., 2010). Supplier relationships are considered more important to gain a competitive advantage for the manufacturer. Large production companies usually have from 2000 to 3000 suppliers. Producers cannot produce organic products unless they cooperate with suppliers. Suppliers must follow buyers' requirements to maintain business relationships. Therefore, we can say that the reluctance of suppliers to move to GSCM is a significant obstacle to the implementation of GSCM in the manufacturing industry.

Government regulations may encourage or discourage the adoption of innovations because the government establishes environmental regulations for the industry. Legal requirements, fees, or time-consuming fees may discourage smaller companies. Tax structures that distort incentives may discourage the industry from implementing GSCM. Government institutions are considered to be obstacles to the development of environmental management in the sense that the institutional process of implementing GSCM is ongoing, but there is minimal institutional support for new ideas for the implementation of GSCM. The government's tendency to encourage old practices is a huge barrier. Therefore, the lack of government support systems is a barrier to the implementation of effective GSCM in Pakistani manufacturing enterprises.

The purpose of this document is to identify various barriers to the implementation of GSCM in manufacturing companies in Pakistan, further identify the contextual relationship between the barriers identified for the implementation of GSCM, classify these barriers according to their driving force and dependence, and finally develop PLS model based on SEM and carried out unreasonable interviews with senior managers of manufacturing companies in Pakistan. Green supply chain management (GSCM) includes traditional supply chain management practices that combine environmental criteria or problems with a purchasing decision of the organization and long-term supplier relationships. Green supply chains aim to reduce waste within the industrial system to save energy and prevent the dispersion of hazardous materials in the environment. Recognizes the disproportionate impact of supply chain processes on the environment in the organization. GSCM is a summary of organic production, ecological distribution, and transport of green with internal environmental management systems and ecological information systems to achieve effective results. GSCM will eliminate or minimize waste in the form of energy, emissions, hazardous waste, chemicals, and solids.

5.2.1.3 Competitive Barriers

Many barriers faced by manufacturing firms of Pakistan during the adoption of green supply chain management practices. We will discuss some critical competitive barriers in the following paragraph for the development of in-depth understanding. GSCM and conventional supply chain management differ in various ways. Green supply chain management takes considerations to ecology as well as the economy as an objective, while conventional supply chain management is usually concentrated on the economy as a single objective of GSCM is ecological, integrated, and ecologically optimized, while conventional supply chain management does not take into account toxicological effects of man. Conventional supply chain management focuses more on controlling the final product, regardless of their harmful effects on the environment during production and distribution. Ecological requirements are vital criteria for products and production, while the company must guarantee its economic balance while maintaining competitiveness and profitability.

A company with a higher quality of human resources, for example, better training or education, will help implement supply chain management. The quality of human resources can provide new ideas for companies, quickly learn new technologies, share knowledge, and use new technologies to solve problems. On the other hand, a weak workforce and the lack of IT specialists are becoming a competitive barrier for companies. However, due to financial constraints, the quality of human resources is a barrier. For this reason, the low quality of human resources is a significant obstacle to the implementation of GSCM in manufacturing companies in Pakistan.

In the current scenario, market uncertainty is very high due to global competitiveness and customer requirements. Global researchers and reference specialists develop and implement strategies. The external environment in which the company develops its business will also affect the innovative ability and intention to innovate. Competition and uncertainty on the market are the most critical barriers to achieving GSCM in the Pakistani manufacturing industry.

Innovative ecological practices are associated with the clarity of ecological practices, the accumulation of knowledge related to the environment, organizational stimulation, and the quality of human resources. Innovative ecological practices include the disposal of hazardous solid waste, energy-saving, reuse, and recycling of materials. Green innovative practices promote innovative design, new market opportunities, and make them better than others. However, due to market competition and costs, organizations are trying to reduce costs. The implementation of GSCM practices is associated initially with significant investments. Financial restrictions also lead to opposition to the implementation of ecological practices (Ravi & Shankar, 2005). From the previous discussion, we hope that the lack of implementation of green practices is the most critical obstacle to the implementation of efficient GSCM in Pakistani manufacturing enterprises. The main obstacle for GSCM observed in manufacturing companies in Pakistan is the lack of knowledge about the benefits of organic products. The client's requirements become the most critical type of external pressure. The customer's conscience means that the customer demands ecological products; the company needs to change technology and organization for innovative organic products. However, in manufacturing companies in Pakistan, due to the lack of knowledge of customers about the benefits of green products, producers produce products that are not organic. In the USA, it is estimated that 75% of consumers say that their purchases are influenced by reputation, and 80% would be willing to pay more for organic products (Lamming & Hamapson, 1996). Therefore, we can say that the lack of knowledge about customers is an essential obstacle to the implementation of GSCM in the Pakistani manufacturing industry.

IT systems support cooperation processes in the supply chain and improve supply chain efficiency. An efficient IT and technology system is very much needed to support GSCM at various stages of the product life cycle. It can be beneficial in product development programs that include environmental design, recovery, and reuse (Ravi & Shankar, 2005). Information support is needed to develop links to achieve effective GSCM in the manufacturing industry. It is necessary to handle information flows related to the forward and backward flow of materials and other resources for the efficient management of green SC. Besides, enabling IT reduces paper consumption, which supports the GSCM philosophy. Then, the lack of IT implementation is an essential obstacle to achieving an efficient GSCM.

Technology is a kind of knowledge. An organization with extensive experience in the application and implementation of related technologies will have a higher capacity for technological innovation. The organization will have a more exceptional innovative ability when knowledge can be more easily shared within the organization. Technological progress can be achieved through greater portability. It is easy to share technology transfer or share technology knowledge (Cooper, 1994). Innovation and technology contain innovations in corporate culture, supporting new ideas, processes, and solutions by all employees of the company (Digalwar & Metri, 2004). The organization's resistance to the adoption of technological progress is resistant to change. An organizational barrier means difficulties in implementing a fundamental change in the organization. This is especially important when changes occur in the main features of the organization, such as organizational goals, forms of power, underlying technology, operational strategy, and market strategy. Therefore, resistance to technological progress is a significant competitive barrier to the implementation of GSCM in manufacturing companies in Pakistan.

Due to the high uncertainty of the market and competition on the international market, it is tough for the industry to reduce costs and implement GSCM at the same time. That is why competition plays a vital role in adopting and not applying GSCM practices. Lack of knowledge about environmentally friendly goods and services in society is an obstacle to the adoption of ecological awareness-raising practices. Lack of knowledge in society means that customers are not yet aware of organic products and their benefits to society. Lack of understanding and knowledge among suppliers is an obstacle to the implementation of GSCM practices. The supplier's failure/unwillingness to switch to GSCM (lack of willingness to exchange information), lack of understanding among stakeholders in the supply chain, etc. are barriers related to suppliers. A company with a higher quality of human resources, for example, better training or education, will help implement supply chain management. The quality of human resources can provide new ideas for companies, quickly learn new technologies, share knowledge, and use new technologies to solve problems. However, due to financial constraints, the quality of human resources is a barrier. For this reason, the low quality of human resources is a significant obstacle to the implementation of GSCM in manufacturing companies in Pakistan.

We can, therefore, say that the flexibility of switching to GSCM is a significant obstacle to the implementation of GSCM in manufacturing companies in Pakistan. The lack of green architects, consultants, green developers, and contractors in the region is the driving force of supply chain management, which harms the rural area, because most architects, consultants, green developers, and contractors are interested in working in urban areas because in the urban area they are more suitable to earn handsome salaries and benefits (Malczewski, 1999). Therefore, more emphasis should be placed on this obstacle to the implementation of green chain management in manufacturing companies in Pakistan. Technology is a kind of knowledge. The organization will have a more exceptional innovative ability when knowledge can be more easily shared within the organization. Therefore, the lack of acceptance of progress in the field of new technologies is an essential obstacle to the implementation of GSCM in manufacturing companies in Pakistan.

The ecological awareness of customers and the increase of image on the market every day indirectly forced manufacturing companies to think about cleaner production thanks to the knowledge of ecological supply chain management practices. Foundries play an essential role in the economy of the country and, therefore, should start adopting green supply chain management as a simultaneous strategy to build their environmental image, although, in most countries, they are still on foot. Barriers that impede the implementation of ecological supply chain management pose a severe challenge to both technical experts and managers of manufacturing companies.

5.2.1.4 Organizational Barriers

Several organizational barriers faced by manufacturing firms of Pakistan during the adoption of green supply chain management practices. We will discuss critical organizational barriers in the following paragraph for the development of clear understanding. Top management support and commitment is necessary for any strategic program success. Lack of top management commitment is one of the barriers to implementation of green supply chain management in the context of manufacturing firms of Pakistan.

There is a lack of education and knowledge necessary to adopt greener practices in the supply chain in manufacturing companies in Pakistan. The literature review also concluded that a lack of knowledge is a significant obstacle to the implementation of

GSCM. Lack of knowledge is the current barrier for Pakistani manufacturing companies. Organizations and individuals do not understand GSCM practices and the specific benefits associated with them.

The size of the industry can be a big obstacle. One of the participants stressed that Pakistani manufacturing companies are still relatively small and only a few large industries. The priority for the government is to encourage producers to stimulate the growth and development of industry and the country. Unfortunately, this leads to a less rigorous approach of the government to the issue of environmental protection. "Pakistan is still young and childish, and heavy industry is not as important as China and European countries, so it can still be managed and should be somewhat open to foreign investors." Besides, the manufacturing industry in Pakistan is still small, and, as the government is trying to attract foreign investors, it tends to understand more of the application of security rules and regulations. Future research can be conducted on the relationship between industry dimensions and the implementation of GSCM practices. Informal links and improved communication help organizations adopt Green's practices (Yu Lin & Hui Ho, 2008). Training and education are the main requirements to succeed in implementing GSCM in any organization. Management can encourage employees to learn ecological information. Organizations can provide rewards for green employees. Employees can get help when they face ecological problems and can get help in learning green information.

Government regulations may encourage or discourage the adoption of innovations because the government establishes environmental regulations for the industry. Legal requirements, fees, or time-consuming fees may discourage smaller companies. Tax structures that distort incentives may discourage the industry from implementing GSCM. Government institutions are considered to be obstacles to the development of environmental management in the sense that the institutional process of implementing GSCM is ongoing, but there is minimal institutional support for new ideas for the implementation of GSCM. The government's tendency to encourage old practices is a huge barrier. Therefore, the lack of government support systems is a barrier to the implementation of effective GSCM in Pakistani manufacturing enterprises. Support and commitment of top management are essential for the success of any strategic program (Hamel & Prahalad, 1989; Zhu & Sarkis, 2007). Support from senior management is particularly useful for environmental practices such as GSCM. The top management has a significant ability to influence and support the actual training and implementation of environmental initiatives throughout the organization (Khan & Dong, 2017a, b). Senior management provides ongoing support to GSCM in strategic plans and action plans for their successful implementation. Lack of commitment from senior management is one of the obstacles to implementing GSCM in manufacturing companies in Pakistan.

The government can act as a driving force and barrier. Almost all participants noticed that it is necessary to change the vision of the government in terms of environmental aspects. Seven participants mentioned that the government is the driving force behind the implementation of green practices in the supply chain, although it is still in the child's development phase. Routroy (2009) argues that governments should participate in activities aimed at educating and promoting

GSCM practices. Participants mentioned that the government has recently established meetings with producers to discuss greener practices. It is a factor that contributes to the awareness and significance of GSCM. The Government of Pakistan is seeking to enforce a limited number of environmental laws. "We have meetings, and the government says, please take into account, do things in this way and grant a period of completion, but it is not yet in the mandatory phase" (senior executive). This is in line with the literature review, as the researchers suggest that governments have a strong driving force in implementing GSCM practices. The current lack of government support means that the government does not provide policies that benefit the industry, nor does it grant unique benefits to organizations that implement GSCM practices (Khan, Dong, Zhang, et al., 2017a, b). Respondents often mentioned the lack of legislation and incentives to motivate the organization. Two participants mentioned that the government is slowly beginning to see the need and importance of green supply chains. However, these are the first days in the country, and we still have much work to do. Six of the eight participants complained about the lack of incentives. These participants suggested that if incentives were established, the barrier would become a driving force. This lack of support is an excellent inhibitor, and several organizations are fighting for incentives to start a business (Khan, Qianli, et al., 2017; Khan, Zhang, et al., 2018). "We are fighting for government incentives to ensure that our competitors and we are involved in initiatives related to green supply chains, we have an association that fights for tax breaks for companies that implement them" (general director of supply chain). In the case of Pakistan, the lack of government regulations and support has been listed as one of the biggest obstacles to the implementation of GSCM.

Legal requirements, fees, or time-consuming may discourage smaller companies. Tax structures that distort incentives may discourage the industry from implementing GSCM. Government institutions are considered obstacles to the development of environmental management in the sense that the institutional process of implementing supply chain management is underway, but minimal institutional support is provided for new ideas for the implementation of green supply chain management. Therefore, the lack of government support systems is a barrier to the implementation of effective GSCM in Pakistani manufacturing enterprises. Informal links and improved communication help organizations adopt Green's practices. Training and education are the main requirements to achieve the successful implementation of green supply chain management in enterprises. Management can encourage employees to learn ecological information. Organizations can provide rewards for green employees. Employees can get help when they face ecological problems and can get help in learning environmental information. Innovative ecological practices include the disposal of hazardous solid waste, energy-saving, reuse, and recycling of materials. Green innovative practices promote innovative design, new market opportunities, and make them better than others. However, due to market competition and costs, organizations are trying to reduce costs.

The GSCM field can be said to be at an early stage of development, both in academic and practical terms. In the academic aspect, in order to successfully and empirically develop in theory in this field, some scales of measurement of many useful and verifiable elements are needed. Using the literature in the supply chain (operations) and environmental management, we present a series of scales that can be used to evaluate practices in this field. Practically, organizations can also benefit from the development of reliable and valid scales to measure the implementation of GSCM practices. Specialists can use these scales to perform benchmarking tests, continuous improvement, and project management activities while trying to implement GSCM practices (Khan, Dong, Zhang, 2016, 2017, 2018). The contribution of this study is to help manufacturers understand different aspects of implementing GSCM practices and identify the strengths and weaknesses of implementing their GSCM practices. GSCM practices the scale of implementation measurement; we present an analysis based on the empirical study of manufacturing companies in Pakistan.

To some extent, they are very advanced in practice and technology due to their relationships with international partners and the community. This study aims to examine the barriers and factors that stimulate GSCM in the context of Pakistani manufacturing companies (Khan, Dong, SongBo, et al., 2017a, b; Khan, Zaman, Zhang, 2016). Green supply chain management (GSCM) is known as a method of improving the recital of processes and products under the needs of ecological principles. In previous paragraphs, several environmental barriers to supply chain management in Pakistani manufacturing companies and significant barriers were analyzed. The elimination of these barriers will help to implement green supply chain management in manufacturing companies in Pakistan.

5.2.2 GSCM Drivers in Manufacturing Firms of Pakistan

From the perspective of Pakistan, many firms are adopting green supply chain management practices in their different activities and operations. Undeniably, during interviews with senior management of manufacturing firms of Pakistan, we have identified several drivers and motivating factors, which not only drives firms towards green supply chain practices but also enforce and create pressure on firms through different perspectives to implement green practices in their end-to-end supply chain operations for environmental sustainability (Zhang et al., 2018a, b; Golpîra et al., 2018; Khan, 2019). Most firms are not adopting green practices for environmental sustainability, but they aim to enhance their financial performance and organizational performance and gain competitive advantages. The drivers of GSCM in the manufacturing firms of Pakistan mainly have been divided in the following categories for better understanding.

5.2.2.1 Economic Drivers

Mainly green supply chain practices are implemented by firms to enhance their financial performance in terms of reduction in waste, recycling, re-manufacturing, and renewable energy sources and also reduce their reverse logistics cost and financial penalties from regulatory authorities (Golpîra and Khan, 2019; Khan et al., 2019; Khan & Zhang, 2019).

Green supply chain management (GSCM) aims to reduce waste and pollution by integrating ecological thinking with product design and end-of-life management. The implementation of these chains depends on the controllers. Drivers are defined as forces that motivate organizations to implement eco-supply chains. It is necessary to understand the various factors that interfere in the management of the ecological supply chain, especially in the manufacturing industries, which aim to minimize the amount of waste and pollution. Organizations participating in green supply chain practices benefit from reduced costs and increased benefits, which makes these practices useful in developing sustainable competitive advantages (Routroy, 2009). In cooperation with supply chain partners, suppliers, and customers, manufacturers can have a better strategic and competitive position if they optimize environmental management. Joint organizational activities to reduce the environmental burden associated with products and processes can help reduce unnecessary waste and improve the efficiency of the supply chain. Drivers who made companies aware of the importance of managing an environment-friendly supply chain are the intensity of competition, changes in government policy, and customer awareness. Rao and Holt (2005) confirmed that there is a significant relationship between drivers, economic performance, and competitiveness in implementing eco-supply chains in a sample of enterprises in Southeast Asia.

Green supply chain management is the process of using organic raw materials from the supply chain and transforming these data into products that can be recovered and reused after the end of the life cycle, thus creating a sustainable supply chain. In the context of sustainable development, this will allow us to preserve more resources for future generations, and at the same time, protect our natural environment, which will result in a better and safer world. The greening of supply chains can increase the competitive advantage and economic performance of an organization. "Production organizations involved in partnerships for an ecological project with their suppliers and customers can develop organizational capabilities that will be reflected not only in environmental performance but also in other dimensions of performance, such as cost and quality." This may result in lower costs, improved operational efficiency, greater flexibility, improved sales, improved value for the customer, and improved social image. However, it has also been shown that the management of an ecological supply chain requires a high initial investment. This may include the adoption of advanced technologies and the hiring, motivating, and training of good quality employees. It is crucial to integrate the concept of environmental management along the supply chain in order to maximize the benefits of greener supply chains and the competitive advantages it provides. Ultimately, it increases the benefits and objectives of market share.

The rule of law is a system of government in which society adopts a set of good, just, and just laws in which society and its government govern. The government's legal regulations define the rules of the game, and organizations are required to comply with regional, national, and international regulations and, at the same time, meet the client's requirements. It is claimed that regulatory pressure is one of the main drivers of the implementation of green supply chains. Government regulations can be the driving force behind the implementation of green supply chains. Governments have the right to act as drivers in the implementation of green supply chains.

Legislation and government regulations act as incentives for organizations to reduce the use of nonrenewable resources and reduce greenhouse gas emissions. An environmental policy focused directly on emissions is the most critical element of a cost-effective environmental policy strategy. The regulations increase the risk of a fine for noncompliance between companies. Global environmental requirements have gone from pollution control in production plants to the actual life cycle of products. This increases the scope of government participation in initiatives related to green supply chains.

The demand for eco-friendly products and services by customers is considered one of the most critical factors driving green initiatives. Customers are increasingly aware of the negative impact of some commercial activities on the environment. The public and clients are putting pressure on organizations to implement greener supply chains, ultimately reducing the environmental damage, they cause. Customers take the environment into account when making purchasing decisions. Customers expect organic products that require the ISO 14001 certificate.

Undeniably, the media play an essential role in informing and educating the public about environmental degradation; customers are increasingly concerned about environmental issues. This can cause the public to buy fewer products that they perceive as harmful to the environment. Customers need not only the right quality products but also require organic products. This encourages organizations to follow the implementation of green supply chain management practices and affect the entire production process, from design to the end of the product's life.

Economic benefits may be the driving force of ecological supply chain management and may reduce unit costs of manufactured products or services without affecting the intended use or quality of the product (Bhool & Narwal, 2013). It should be remembered that cost reductions should not result in lowering the quality of the products. Elimination of ingredients that do not add value to products can lead to cost reduction. Also, economic benefits can be obtained by using less energy, water, and raw materials in the production process, which not only protects the environment but also reduces production costs. The implementation of ecological supply chain management is widely considered to be a win-win situation, reducing costs, improving operational efficiency, and improving the reputation of the organization. Benefits such as lowering costs, increasing customer satisfaction, new market opportunities, improving the company's image, and higher profits can be achieved by implementing green supply chain management practices. Eco-management practices in the supply chain offer the potential to generate competitiveness and economic performance.

In organizations, a controller that exists itself is known as an internal controller. There are many different GSCM controllers related to the organization that includes the organization's environmental mission, employee participation/motivation, willingness to reduce costs, and investor. Employer policy/environmental mission, employee participation/motivation, willingness to reduce costs, costs of eliminating harmful materials, etc. are known as an internal GSCM engine. Carter et al. (1998)

stated that the personal and ethical values of the founder of the organization, not only top management but also mid-level managers, are positively linked to environmental purchases. Government regulations and legislation seem to be a strong impulse for GSCM, especially if companies are positive and innovative in their approach to compliance. Zhu et al. (2005) and Beamon (1999) stated in their research that the main driver of GSCM's efforts in enterprises is government regulation and legislation.

Customers play a vital role in managing the green supply chain, and there are many customer problems as a driving force for GSCM practices. GSCM plays a vital role in choosing a green supplier. It was observed that suppliers could help in providing essential ideas used in the implementation of environmental projects, services for customers.

Despite numerous counter-evidence, the myth that the ecological process costs extra money is maintained. Some of the factors responsible for the persistence of this myth are inertia, lack of a systematic approach, and reluctance to engage in sustained and modified thinking that is necessary to create an eco-friendly supply chain. However, the essential benefit of green supply chains is the long-term positive net effect on the organization's financial performance, and this has been proved based on both analysis and empirical evidence. In the context of Pakistan, several manufacturing companies use green supply chain practices in their supply chain operations to improve financial and environmental benefits with ethical issues.

5.2.2.2 Environmental Drivers

Legislation and government regulations act as incentives for organizations to reduce the use of nonrenewable resources and reduce greenhouse gas emissions. An environmental policy focused directly on emissions is the essential element of a cost-effective environmental policy strategy. The regulations increase the risk of a fine for noncompliance between companies. The regulations increase the costs of organizations that do not implement green practices through sanctions and fines. Global environmental requirements have gone from pollution control in production plants to the actual life cycle of products. This increases the scope of government participation in initiatives related to green supply chains. Lee (2008) notes that governments can encourage organizations by encouraging tax cuts and developing infrastructure for environmentally friendly complexes. This demonstrates the power that governments can exert as the driving force behind the implementation of green supply chains.

Undeniably, the implementation of top management plays a significant role in the adoption of green supply chain management practices. Companies adopt green practices to improve environmental sustainability. Besides, employee pressure and media pressure are factors that are not ignored at all costs. Companies adopt green and ecological practices due to pressure from their employees because employees do not feel proper working for polluted companies.

On the other hand, media pressure is also essential. In the context of Pakistan, media freedom is sturdy, and no one can control the media. The government does not play a role in the media and is entirely far from the media.

It is recommended that all industrial waste is removed from some GMP steam traps, whether the waste will be neutralized, meaning that the waste will not become hazardous. This will prevent damage to the environment. Green supply chain management refers to a variety of activities and initiatives taken by the organization to reduce their impact on the environment. The principles of supply chain management include reverse logistics in the process of planning, implementing, and controlling efficient, effective processing of raw materials, process materials, finished products, and relevant knowledge from the point of consumption to and after the deposition of raw material. The duty of manufacturing companies in Pakistan is to develop reverse logistics networks, and options flow in order to refrain from the dissatisfaction of clients and return of ordered medicines. Today, green supply chain management has become a hot topic among medicines. All pharmaceuticals that use chemicals and their products are dangerous, harmful to the environment. The implementation of ecological supply chain management has become a necessity to prevent damage caused to the environment. Production companies are growing day by day but have a low positive attitude to the environment.

Eco-friendly is the driving force behind GSCM, whose initiative is to assess and take responsibility for the organization's environmental impact and its impact on social welfare (Bhool & Narwal, 2013). In today's market, organizations feel socially and environmentally responsible and engage in corporate social responsibility (CSR). It is a set of rules, codes, or voluntary guidelines initiated and promoted by the organization (Broomhill, 2007). Environmental problems are considered more seriously, and CSR is an effective way for the organization to stand out with their peers. Companies that want to improve their image often advertise their activities to protect the natural environment, to show their involvement of stakeholders and society. Organizations look for strategies such as CSR to improve your brand's reputation and image. The purpose of diversifying and promoting the image of the organization makes CSR a driving force in the implementation of GSCM, looking for an alternative to establishing a green environment by planting trees. The definition and scope of GSCM in the literature differed from green purchases to integrated green supply chains that flow from supplier to producer to customer.

Carbon-based production minimizing harmful environmental impact when selecting products and reusing and being recycled and repaired during ecological production practices create maximum flexibility in production processes and recycling. Refinished products not only save money but also improve the long-term ecological balance. The theory of ecological supply chain management is mainly based on ecological production and supply chain technology. The goal is to reduce the negative impact on the environment and improve the primary purpose of the company through supply chain operations, management strategies, and cooperation partnerships. Green supply chain management offers companies the opportunity to improve the performance and image of their companies in general.

The Pakistani economy has proliferated in recent decades. The issue of managing green supply chains has been turned to production and research practice in Pakistan, which is why GSCM is research on fast growth. He deals with environmental problems and efficient use of resources throughout supply chain management, which distinguishes him from managing the traditional supply chain. Companies are trying to reduce costs by greening supply chain management, which also builds a competitive advantage for the firms. On the other hand, the companies are taking advantage of the image of the product by green labeling. The manufacturing companies are facing a more strict environmental emission norm, which forces the Pakistan Environmental Protection Agency (EPA) to reduce harmful effects on environmental sustainability.

5.2.2.3 Competitive Drivers

Green supply chain management adoption by several firms in Pakistan for creating a competitive advantage. Undeniably, companies implement green practices in their supply chain operations to gain a competitive advantage in terms of positive image and reputation of firms, higher quality of products, enhance market power, market share, and size. Besides, green supply chain practices provide an opportunity for firms to export their products all around the globe, because several countries do not accept/allow polluted products in their territory boundaries.

GSCM practices can improve the sense of loyalty towards an organization, which in turn translates into an increase in market share, as has been observed in many cases. Sustainable supply chain management suggests that GSCM is directly related to the goals of enterprises to achieve a larger market share. GSCM practices and higher concentration on R (i.e., reduction, reuse, resale, repair, recycling, reconditioning, and reverse logistics) can help the organization achieve its goals (i.e., profitability), while at the same time market share also increases due to improving customer satisfaction and ecological experience. Recent research also indicates that those organizations that have adopted GSCM as their corporate strategy have improved their results.

Technology is a crucial element of cooperation and cost reduction – slow, inefficient, and insufficient communication results from a lack of technical infrastructure. Pakistan does not have the adequate technological infrastructure. For several decades, companies in Pakistan have been using highly advanced technology to control the damaging impact of supply chain operations on environmental sustainability and their image and reputation.

Support from senior management is particularly useful for environmental practices such as GSCM. Based on the discussion with the interviewees, the organization's management and the top management of the production sector have committed to implementing GSCM practices because they put pressure on companies to comply with these practices. Consider the following quote from the participant: "Green initiatives come from the management and owners of the company because they understand more Western, Oriental and African vision ... and have a wider mentality." Top management is a driver and plays an important role. In Pakistan, support from senior management is the driving force for the organization because they tend to follow GSCM practices because of internal pressure from the board and senior management. Besides, studies should be carried out to understand the causes of the same factor, which in some cases is the controller and, in others, barrier.

5.2.2.4 Organizational Drivers

Organization strategy is also one of the critical drivers for implementing GSCM. Environmental innovation practices lead to improve a firm's environmental performance and indirectly improve financial performance. The finding indicated that firms should change the attitude and perception of intent on environment improvement as an economic and competitive opportunity. Global organizations are exposed to very high competitive pressure. In order to keep customers and market share, companies must initiate and differentiate themselves from competitors, so competitors are also one of the factors. A more ecological supply chain generates significant values in terms of competitiveness and economic performance. The cost reduction aims to save production costs and ultimately increase the profit margin. Benefits include the reduction of direct costs to facilitate the development of cooperation with suppliers and a holistic view of the management supply chain. Other criteria are the low-priority criteria for the manufacturing industry in Pakistan; these are social entities/interested parties, reverse logistics, and supplier. Stakeholders are individuals or groups that influence the company's operations and influence the company involved in environmental practices. The organizational environmental plan is positively influenced by customer pressure, shareholder pressure, regulatory pressure from the government, and pressure from the neighbors and communities. Reverse logistics includes the handling and disposal of returned products and the use of related materials and information. There is evidence that government regulation, legislation, and reverse logistics are essential factors to achieve collaboration between product designers and suppliers to reduce and eliminate product impacts on the environment. Suppliers who pay attention to environmental issues should be able to support companies to be GSCM. However, relationships between companies and suppliers are significant to support ecological implementation.

The role of suppliers is considered an essential element in promoting environmental practices. Manufacturers work with suppliers to set environmental standards to examine eco-supply chain management, such as product design, which can eliminate the environmental impact of production and the development of environmental packaging. Government regulations and regulations are an essential driver of environmental enterprises' efforts to implement GSCM practices. The government of each country has different rules in the manufacturing industry. That is why manufacturing companies need to study environmental standards and standards to ensure that each country maintains market shares. In Pakistan, manufacturing companies must comply with environmental regulations on domestic and foreign markets to meet customer requirements. Regulations and environmental policy encourage organizations to become more environmentally aware. To improve the company's competitiveness, the organization must strive to comply with the strictest environmental protection regulations, both now and in the future. Besides, environmental policy can foster competition in the industry and improve profitability and efficiency in the field of environmental protection.

Market factors affect the decision on the management of the organic supply chain in the manufacturing industry, as most companies are international companies. Manufacturers must comply with the environmental standards and requirements of each country, such as RoHS, WEEE, and EPA. As a result, the manufacturing industry must implement GSCM to meet the export requirements of customers on the international market. Consumer behavior is more prone to pay attention to organic products. They probably buy an environmentally friendly product. The growing number of organizations that have adopted an organic product also points to the need for an appropriate segmentation and positioning strategy.

Social responsibility has become an important variable to create a global competitive advantage. Manufacturers develop an eco-friendly product to meet market requirements to remain competitive on the market. Deterioration of the environment in the last decade has increased the social awareness of environmental problems. The social company has demanded products from a company that conducts ecological activities. Social pressure means that most companies review their environmental practices within their supply chain. Threats or disturbances caused by ecological awareness of the society may create opportunities for companies to reach new customers who want organic products.

Support for senior management is initially a major problem related to the implementation of GSCM. The involvement of senior management plays a vital role in implementing GSCM's strategy to achieve successful results. For example, top management determines the principles of vision, mission, and organization and delegates these principles to employees at various levels who can achieve collaboration across the company. Consumers are more aware of environmental standards and prefer to buy products from companies offering organic products. For this reason, many companies had to adapt to the green strategy to improve the brand image and maintain market share. The ultimate goal of companies is to generate profits for companies. By accepting GSCM, you save production costs and increase profits. Lowering costs refers to the conservation of energy, water, and raw materials. The cost reduction is the improvement of the environment and the reduction of costs. Besides, previous research has confirmed that accepting GSCM can reduce costs. The goal of GSCM is to produce products that respect the environment with minimal resources (materials, energy, water, etc.) and minimal waste. The benefit of GSCM is to create benefits for the organization from a broader perspective.

Finally, many manufacturing firms have been adopted green supply chain practices in Pakistan due to support and motivating factors of lower rejection rate, organizational structure and strategy, government and regulatory authority support, green organizational culture, awareness of customers regarding green products, media and employees' pressure, and improvement in firms' performance.

References

Aidt, T. (2009). Corruption, institutions, and economic development. *Oxford Review of Economic Policy, 25*(2), 271–291.

AlKhidir, T., & Zailani, S. (2009). Going green in supply chain towards environmental sustainability. *Global Journal of Environmental Research, 3*(3), 246–251.

Azevedo et al. (2011). The influence of green practices on supply chain performance: A case study approach, *Transportation Research Part E: Logistics and Transportation Review, 47*(6), 850–871.

Beamon, B. (1999). Measuring supply chain performance. *International Journal of Operations & Production Management, 19*(3), 275–292.

Ben Moussa, F. Z., Rasovska, I., Dubois, S. S. S. S., De Guio, R., & Benmoussa, R. (2016). Reviewing the use of the theory of inventive problem solving (TRIZ) in green supply chain problems. *Journal of Cleaner Production, 142*, 2677–2692. https://doi.org/10.1016/j.jclepro.2016.11.008

Bhool & Narwal (2013). An analysis of drivers affecting the implementation of green supply chain management for the Indian manufacturing industries. *International Journal of Research in Engineering and Technology, 2*(11), 242–254.

Broomhill, R. (2007). Corporate Social Responsibility: Key Issues and Debates, Dunstan Paper No. 1/2007. http://www.som.cranfield.ac.uk/som/dinamic-content/research/doughty/Dunstan1.pdf - [Accessed 22nd Mar 2019].

Carter, C. R., Ellram, L. M., & Ready, K. J. (2006). Environmental purchasing: benchmarking our german counterparts. *International Journal of Purchasing & Materials Management, 34*(4):28–38.

Carter, C. R., Kale, R., Grimm, C. M. (2000). Environmental purchasing and firm performance: an empirical investigation, *Transportation Research Part E: Logistics and Transportation Review, 36*(3), 219–228.

Cooper, J. (1994). *Green logistics, European logistics: Markets, management, and strategy*. Oxford: Blackwell Business.

Davies, A., Thomas, A. J., & John, E. G. (2013). Minimising transportation in manufacturing supply chains. *International Journal of Sustainable Engineering, 7*(2), 144–153. https://doi.org/10.1080/19397038.2013.801531

De Giovanni, P., & Esposito, V. (2012). Covariance versus component-based estimations of performance in green supply chain management. *International Journal of Production Economics, 135*(2), 907–916. https://doi.org/10.1016/j.ijpe.2011.11.001

De Medeiros, J. F., Ribeiro, J. L. D., & Cortimiglia, M. N. (2016). Influence of perceived value on purchasing decisions of green products in Brazil. *Journal of Cleaner Production, 110*, 158–169. https://doi.org/10.1016/j.jclepro.2015.07.100

Digalwar, A. K., & Metri, B. A. (2004). Performance measurement framework for world class manufacturing. *International Journal of Applied Management and Technology, 3*(2), 83–101. ISSN: 1554-4740(www.ijamt.org)

Dongmin, X., Dingyi, Youfang, H., & Bin, Y. (2016). A study of cost and emissions considering speed optimization in trans eurasia transportation based on intelligent transportation. *International Journal on Smart Sensing and Intelligent Systems, 9*(1), 191–214.

Geffen & Rothenberg's (2000). Suppliers and the automotive painting process', *International Journal of Operations and Production Management, 20*(2), 166–86.

Geng, Y., Ma, Z., Xue, B., Ren, W., Liu, Z., & Fujita, T. (2013). Co-benefit evaluation for urban public transportation sector – A case of Shenyang, China. *Journal of Cleaner Production, 58*, 82–91. https://doi.org/10.1016/j.jclepro.2013.06.034

Golpîra, H., & Khan, S. A. R. (2019). A multi-objective risk-based robust optimization approach to energy management in smart residential buildings under combined demand and supply uncertainty. *Energy, 170*(3), 1113–1129.

Golpîra, H., Khan, S. A. R., & Zhang, Y. (2018). Robust smart energy-efficient production planning for a comprehensive job-shop manufacturing system under combined demand and supply uncertainty in the presence of grid-connected microgrid. *Journal of Cleaner Production, 202*(11), 649–665.

Govindan, K., Khodaverdi, R., & Vafadarnikjoo, A. (2015). Intuitionistic fuzzy based DEMATEL method for developing green practices and performances in a green supply chain. *Expert Systems with Applications, 42*(20), 7207–7220. https://doi.org/10.1016/j.eswa.2015.04.030

Green, K. W., Zelbst, P. J., Bhadauria, V. S., & Meacham, J. (2012). Do environmental collaboration and monitoring enhance organizational performance?. *Industrial Management & Data Systems, 112*(2), 186–205.

Grønhaug & Kaufmann, G. (1998). Innovation: A cross-disciplinary perspective. Norwegian University Press, Oslo, Norway.

Gupta, H., & Barua, M. (2017). Supplier selection among SMEs on the basis of their green innovation ability using BWM and fuzzy TOPSIS, *Journal of Cleaner Production, 152*, 242–258.

Hamel, G., & Prahalad, C. K. (1989). "Strategic Intent" Harvard Business Review, May –June, 67(3), 63–78.

Hart, S. L. (1995). A natural-resource-based view of the firm. *Academy of Management Review, 20*: 986–1014.

Hayami, H., Nakamura, M., & Nakamura, A. O. (2015). Economic performance and supply chains: The impact of upstream firms' waste output on downstream firms' performance in Japan. *International Journal of Production Economics, 160*, 47–65. https://doi.org/10.1016/j.ijpe.2014.09.012

Herold, D. M., & Lee, K.-H. (2017). Carbon management in the logistics and transportation sector: An overview and new research directions. *Carbon Management, 8*, 79–97. https://doi.org/10.1080/17583004.2017.1283923

Hsu (2008). Developing an index for online customer satisfaction: Adaptation of American customer satisfaction index, *Expert Systems with Applications, 34*(4), 3033–3042.

Jain, V. K., & Sharma, S. (2014). Drivers are affecting the green supply chain management adaptation: A review. *UP Journal of Operations Management, 13*(1), 54–63.

Jakowczyk, M., Quariguasi, J., Neto, F., Gibson, A., & Van, L. N. (2017). Understanding the market for remanufactured products: What can we learn from online trading and web search sites? *International Journal of Production Research, 55*, 3465. https://doi.org/10.1080/00207543.2016.1275872

Khan, S. A. R. (2019). The nexus between carbon emissions, poverty, economic growth, and logistics operations-empirical evidence from Southeast Asian countries. *Environmental Science and Pollution Research*. https://doi.org/10.1007/s11356-019-04829-4

Khan, S. A. R., Chen, J., Zhang, Y., Golpîra, H., Kumar, A., & Arshian, S. (2019). Environmental, social, and economic growth indicators spur logistics performance: From the perspective of the South Asian Association for Regional Cooperation countries. *Journal of Cleaner Production, 214*(3), 1011–1023.

Khan, S. A. R., & Dong, Q. (2017a). Impact of green supply chain management practices on firms' performance: An empirical study from the perspective of Pakistan. *Environmental Science and Pollution Research, 24*(20), 16829–16844.

Khan, S. A. R., & Dong, Q. (2017b). Does national scale economic and environmental indicators spur logistics performance? Evidence from UK. *Environmental Science and Pollution Research, 24*(34), 26692–26705.

Khan, S. A. R., Dong, Q., SongBo, W., Zaman, K., & Zhang, Y. (2017a). Environmental logistics performance indicators affecting per capita income and sectoral growth: Evidence from a panel of selected global ranked logistics countries. *Environmental Science and Pollution Research, 24*(2), 1518–1531.

Khan, S. A. R., Dong, Q., SongBo, W., Zaman, K., & Zhang, Y. (2017b). Travel and tourism competitiveness index: The impact of air transportation, railways transportation, travel, and transport services on international inbound and outbound tourism. *Journal of Air Transport Management, 58*(1), 125–134.

Khan, S. A. R., Dong, Q., & Zhang, Y. (2016). Research on the measuring performance of green supply chain management: In the perspective of China. *International Journal of Engineering Research in Africa, 27*(12), 167–178.

Khan, S. A. R., Dong, Q., & Zhang, Y. (2017). Role of ABC analysis in the process of efficient order fulfillment: A case study. *Advanced Engineering Forum, 23*(7), 114–121.

Khan, S. A. R., Dong, Q., & Zhang, Y. (2018). The impact of green supply chain practices in business performance: Evidence from Pakistani FMCG firms. *Journal of Advanced Manufacturing Systems, 17*(2), 267–275.

Khan, S. A. R., Dong, Q., & Zhang, Y. (2017). Role of ABC analysis in the process of efficient order fulfillment: A case study. *Advanced Engineering Forum, 23*(7), 114–121.

Khan, S. A. R., Dong, Q., Zhang, Y., & Khan, S. S. (2017a). The impact of green supply chain on enterprise performance: In the perspective of China. *Journal of Advanced Manufacturing Systems, 16*(3), 263–273.

Khan, S. A. R., Dong, Q., Zhang, Y., & Khan, S. S. (2017b). *Research on the decision-making of green reverse logistics in enterprises: A case study on electronic product manufacturers from the perspective of South Africa.* 2nd international conference on advances in management engineering and information technology in Shanghai, during April 23–24, 2017 (AMEIT 2017), ISBN: 978-1-60595-457-8, pp. 11–18.

Khan, S. A. R., Qianli, D., SongBo, W., Zaman, K., & Zhang, Y. (2017). Environmental logistics performance indicators affecting per capita income and sectoral growth: Evidence from a panel of selected global ranked logistics countries. *Environmental Science and Pollution Research, 24*(2), 1518–1531. https://doi.org/10.1007/s11356-016-7916-2

Khan, S. A. R., Zaman, K., & Zhang. (2016). The relationship between energy-resource depletion, climate change, health resources, and the environmental Kuznets curve: Evidence from the panel of selected developed countries. *Renewable and Sustainable Energy Reviews, 62*(9), 468–477.

Khan, S. A. R., & Zhang, Y. (2019). *Strategic supply chain management, EAI/Springer innovations in communication and computing series.* Springer Publisher, ISBN 978-3-030-15058-7, https://www.springer.com/gp/book/9783030150570

Khan, S. A. R., Zhang, Y., Anees, M., Golpîra, H., Lahmar, A., & Dong, Q. (2018). Green supply chain management, economic growth, and environment: A GMM based evidence. *Journal of Cleaner Production, 185*(6), 588–599.

Khiewnavawongsa, S., & Schmidt, E. K. (2013). *Barriers to green supply chain implementation in the electronics industry. Bangkok: IEEE.* 2013 IEEE international conference on industrial engineering and engineering management, Bangkok: Thailand.

Kim, I., & Min, H. (2011). Measuring supply chain efficiency from a green perspective. *Management Research Review, 34*(11), 1169–1189. https://doi.org/10.1108/01409171111178738

Kim, S.-T., & Lee, S.-Y. (2012). Stakeholder pressure and the adoption of environmental logistics practices: Is eco-oriented culture a missing link?, *The International Journal of Logistics Management, 23*, 238–259

Kirschstein, T., & Meisel, F. (2015). GHG-emission models for assessing the eco-friendliness of road and rail freight transports. *Transportation Research Part B: Methodological, 73*, 13–33. https://doi.org/10.1016/j.trb.2014.12.004

Kleindorfer, P. R., Singhal, K, & Van Wassenhove, L. N. (2005). Sustainable operations management. *Production and Operations Management, 14* (4): 482–492.

Kumar, S., Chattopadhyaya, S., & Sharma, V. (2012). Green supply chain management: A case study from Indian electrical and electronics industry. *International Journal of Soft Computing and Engg., 16*, 275–281.

Kushwaha, G. S., & Sharma, N. K. (2016). Green initiatives: A step towards sustainable development and a firm's performance in the automobile industry. *Journal of Cleaner Production. Elseiver Ltd, 121*, 248. https://doi.org/10.1016/j.jclepro.2015.07.072

Kusi-Sarpong, S., Sarkis, J., & Wang, X. (2016). Assessing green supply chain practices in the Ghanaian mining industry: A framework and evaluation. *International Journal of Production Economics, 181*, 325–341. https://doi.org/10.1016/j.ijpe.2016.04.002

Lai, K. H., Wong, C. W. Y., & Cheng, T. C. E. (2012). Ecological modernisation of Chinese export manufacturing via green logistics management and its regional implications. *Technological Forecasting and Social Change, 79*(4), 766–770. https://doi.org/10.1016/j.techfore.2011.10.004
Lai, K., & Wong, C. W. Y. (2012). Green logistics management and performance: some empirical evidence from Chinese manufacturing exporters. *Omega 40*(3), 267–282.
Lamming, R., & Hamapson, J. (1996). The environmental as a Supply Chain Management issue. *British Journal of Management, 7*((March Special Issue)), 45–62.
Lau, K. H. (2011). Benchmarking green logistics performance with a composite index. *Benchmarking: An International Journal, 18*(6), 873–896. https://doi.org/10.1108/14635771111180743
Lee, K. (2008). Opportunities for green marketing: young consumers, *Marketing Intelligence & Planning, 26*(6), 573–586.
Liotta, G., Stecca, G., & Kaihara, T. (2015). Optimisation of freight flows and sourcing in sustainable production and transportation networks. *International Journal of Production Economics, 164*, 351–365. https://doi.org/10.1016/j.ijpe.2014.12.016
Liu, L., Tang, M., & Xue, F. (2012). The impact of manufacturing firms' green supply chain management on competitive advantage. *Advanced Materials Research, 475*, 3349–3354. https://doi.org/10.4028/www.scientific.net/AMR.472-475.3349
Luthra, S., Govindan, K., Kannan, D., Mangla, S. K., & Garg, C. P. (2017). An integrated framework for sustainable supplier selection and evaluation in supply chains. *Journal of Cleaner Production, 140*, 1686–1698. https://doi.org/10.1016/j.jclepro.2016.09.078
Luthra, S., Kumar, V., Kumar, S., & Haleem, A. (2011). Barrier to implement green supply chain management in the automobile industry using interpretive structural modeling technique-an Indian perspective. *Journal of Industrial Engineering and Management, 4*(2), 231–257.
Malczewski, J. (1999). *GIS and multi-criteria decision analysis*. New York: Willey.
Marchet, G., Melacini, M., & Perotti, S. (2013). Environmental sustainability in logistics and freight transportation: A literature review and research agenda. *Journal of Manufacturing Technology Management, 25*(6), 775–811. https://doi.org/10.1108/JMTM-06-2012-0065
Massoud, M. A., Fayad, R., Kamleh, R., & El-Fadel, M. (2010). Environmental management system (ISO 14001) certification in developing countries: Challenges and implementation strategies 1. *Environmental Science & Technology, 44*(6), 1884–1887.
Mirhedayatian, S. M., Azadi, M., & Farzipoor Saen, R. (2014). A novel network data envelopment analysis model for evaluating green supply chain management. *International Journal of Production Economics, 147*(PART B), 544–554. https://doi.org/10.1016/j.ijpe.2013.02.009
Mitra, S., & Datta, P. P. (2014). Adoption of green supply chain management practices and their impact on performance: an exploratory study of Indian manufacturing firms, *International Journal of Production Research, 52*(7), 2085–2107.
Mudgal, R. K., Shankar, R., Talib, P., & Tilak, R. (2009). Greening the supply chain practices: An Indian perspective of enablers relationships. *International Journal of Advanced Operations Management, 1*, 151.
Mudgal, V., Madaan, N., Mudgal, A., Singh, R. B., & Mishra, S. (2010). Effect of toxic metals on human health. *The Open Nutraceuticals Journal, 3*, 94–99.
Nie, X. (2016). Research on the management innovation in the green supply chain: An empirical analysis based on probit model. *International Journal of Smart Home, 10*(2), 153–164. https://doi.org/10.14257/ijsh.2016.10.2.15
Porter & Linde (1995). Toward a new conception of the environment-competitiveness relationship. *The Journal of Economic Perspectives, 9*(4), 97–118.
Preuss, (2001). In dirty chains? purchasing and greener manufacturing. *Journal of Business Ethics, 34*, 345–359.
Rao, P., & Holt, D. (2005). Do green supply chains lead to competitiveness and economic performance? *International Journal of Operations & Production Management, 25*(9), 898–916. https://doi.org/10.1108/01443570510613956.
Ravi, V., & Shankar, R. (2005). Analysis of interactions among the barriers of reverse logistics. *International Journal of Technological Forecasting & Social Change, 72*(8), 1011–1029.

Reimann, M., Schilke, O., & Thomas, J. S. (2009). Customer relationship management and firm performance: The mediating role of business strategy. *Journal of the Academy of Marketing Science, 38*(3), 326–346. https://doi.org/10.1007/s11747-009-0164-y

Rostamzadeh, R., Govindan, K., Esmaeili, A., & Sabaghi, M. (2015). Application of fuzzy VIKOR for evaluation of green supply chain management practices. *Ecological Indicators, 49*, 188–203. https://doi.org/10.1016/j.ecolind.2014.09.045

Routroy, S. (2009). Antecedents and drivers of green supply chain management implementation in a manufacturing environment. *Journal of Supply Chain Management, 6*(1), 20–35.

Schenkel, M., Caniëls, M. C. J., Krikke, H., & Van Der Laan, E. (2015). Understanding value creation in closed-loop supply chains – Past findings and future directions. *Journal of Manufacturing Systems, 37*, 729–745. https://doi.org/10.1016/j.jmsy.2015.04.009

Schniederjans, D. G., & Hales, D. N. (2016). Cloud computing and its impact on economic and environmental performance: A transaction cost economics perspective. *Decision Support Systems, 86*, 73–82. https://doi.org/10.1016/j.dss.2016.03.009

Scur, G., & Barbosa, M. E. (2017). Green supply chain management practices: Multiple case studies in the Brazilian home appliance industry. *Journal of Cleaner Production, 141*, 1293–1302. https://doi.org/10.1016/j.jclepro.2016.09.158

Sheu, J. B., & Chen, Y. J. (2014). Transportation and economies of scale in recycling low-value materials. *Transportation Research Part B: Methodological, 65*, 65–76. https://doi.org/10.1016/j.trb.2014.03.008

Simpson, D. (2012). Knowledge resources as a mediator of the relationship between recycling pressures and environmental performance. *Journal of Cleaner Production, 22*(1), 32–41. https://doi.org/10.1016/j.jclepro.2011.09.025

Taylor, P., Mitra, S., & Datta, P. P. (2014). Adoption of green supply chain management practices and their impact on performance: An exploratory study of Indian manufacturing firms, (May), 37–41 . https://doi.org/10.1080/00207543.2013.849014

Tseng & Chiu (2013). Evaluating firm's green supply chain management in linguistic preferences, *Journal of Cleaner Production, 40*, 22–31.

Walker, H., Di Sisto, L., & McBain, D. (2008). Drivers and barriers to environmental supply chain management practices: Lessons from the public and private sectors. *Journal of Purchasing and Supply Management, 14*(1), 69–85.

Wang, Z., Subramanian, N., Gunasekaran, A., Abdulrahman, M. D., & Liu, C. (2015). Composite sustainable manufacturing practice and performance framework: Chinese auto-parts suppliers' perspective. *International Journal of Production Economics, 170*, 219–233. https://doi.org/10.1016/j.ijpe.2015.09.035

Wee, Y. S., & Quazi, H. A. (2005). Development and validation of critical factors of environmental management. *Industrial Management & Data Systems, 105*(1), 96–114. https://doi.org/10.1108/02635570510575216.

Woo, C., Kim, M. G., Chung, Y., & Rho, J. J. (2016). Suppliers' communication capability and external green integration for green and financial performance in the Korean construction industry. *Journal of Cleaner Production, 112*, 483–493. https://doi.org/10.1016/j.jclepro.2015.05.119

Yang, C., Lu, C., Jing, J., & Bernard, P. (2013). The effect of green supply chain management on green performance and firm competitiveness in the context of container shipping in Taiwan. *Transportation Research Part E, 55*, 55–73. https://doi.org/10.1016/j.tre.2013.03.005

Yu Lin, C., & Hui Ho, Y. (2008). An Empirical Study on Logistics services provider, intention to adopt Green Innovations. *Journal of Technology, Management and Innovation, 3*(1), 17–26.

Yu, F., Xue, L., Sun, C., & Zhang, C. (2016). Product transportation distance based supplier selection in a sustainable supply chain network. *Journal of Cleaner Production, 137*, 29–39. https://doi.org/10.1016/j.jclepro.2016.07.046

Zailani, S. H. M., Eltayeb, T. K., Hsu, C. C., & Tan, K. C. (2012). The impact of external institutional drivers and internal strategy on environmental performance. *International Journal of Operations & Production Management, 32*(6), 721–745.

Zhang, Y., Golpîra, H., & Khan, S. A. R. (2018a). The relationship between green supply chain performance, energy demand, economic growth, and environmental sustainability: Empirical evidence from developed countries. *LogForum Scientific Journal of Logistics, 14*(4), 479–494. https://doi.org/10.17270/J.LOG.2018.304

Zhang, Y., Golpîra, H., & Khan, S. A. R. (2018b). The impact of GSCM on manufacturing enterprise's performance. *Journal of Advanced Manufacturing Systems, 17*(4), 445–459.

Zhou, M., Pan, Y., Chen, Z., & Li, B. (2015). Enterprise behavior under Cap-and-Trade conditions: An experimental study with dynamic system models. *Journal of Simulation, 10*, 12–23. https://doi.org/10.1057/jos.2014.36

Zhou, Y., Gong, D. C., Huang, B., & Peters, B. A. (2017). The Impacts of Carbon Tariff on Green Supply Chain Design. *IEEE Trans. Autom. Sci. Eng, 14*, 1542–1555.

Zhu & Sarkis (2004). Relationships between operational practices and performance among early adopters of green supply chain management practices in Chinese manufacturing enterprises. *Journal of Operations Management, 22*(3), 265–289.

Zhu, Q., & Sarkis, J. (2007). The moderating effects of institutional pressures on emergent green supply chain practices and performance. *International Journal of Production Research, 45*, (18–19), 4333–4355.

Zhu, W., Chew, I. K. H., & Spangler, W. (2005). CEO Transformational leadership and organizational outcomes: The mediating role of human-capital-enhancing human resource management. *The Leadership Quarterly, 16*(1), 39–52.

Zhu, Q., Sarkis, J., & Lai, K. H. (2008). Confirmation of a measurement model for green supply chain management practices implementation. *International Journal of Production Economics, 111*(2), 261–273. https://doi.org/10.1016/j.ijpe.2006.11.029

Chapter 6
Practical Implications and Recommendations for Green Supply Chain Management

6.1 Key Findings from Mixed Approached

Green supply chain management practices and their effect on the Pakistani manufacturing firms have been discussed in previous chapters with necessary statistical analysis. In this book, the researcher has used two different research methodologies including the qualitative and quantitative to optimize the benefits of qualitative and quantitative techniques. First of all, the researcher collected data from different Pakistani manufacturers, and during the collection of data, the researcher also conducted structural interviews with senior management of the firms including supply chain director, supply chain manager, and supply chain senior manager. After the collection of data, the researcher used different statistical techniques to convert data into the useable form to test the research hypothesis.

6.2 Practical Implications

In this thesis, the researcher covers green supply chain management practices' effect on organizational performance. For this purpose, the researcher develops a formal and detailed model (Fig. 4.12). In the first layer is internal environmental management and green information systems; the second layer is covering green distribution, ecological product design, green manufacturing, and green transportation; and finally, the third layer includes competitive advantage, the economic performance of the firms, and environmental performance, which have direct and significant effects on organizational performance.

The results reveal that internal environmental management and green information systems have a positive and significant effect on the ecological design, green

S. A. R. Khan, *The Critical Success Factors of Green Supply Chain Management in Emerging Economies*, EAI/Springer Innovations in Communication and Computing, https://doi.org/10.1007/978-3-030-42742-9_6

manufacturing, green transportation, and green distribution. Besides, competitive advantage, economic performance, and environmental performance also have a positive association with green practices (ecological design, green manufacturing, green transportation, and green distribution). Finally, the results indicate that environmental performance, competitive advantage, and economic performance of the firms have direct and positive effects on the organizational performance (Ref. Chaps. 4 and 5).

6.2.1 The Key Barriers

The findings confirm that there are many barriers and hurdles confronted by Pakistani manufacturing firms during the adoption of green practices, while some of those barriers are not easy to overcome without the proper, sufficient support and involvement of senior management. In the context of Pakistani manufacturing firms, the key barriers are lack of senior management support, lack of skilled labors, low ability to control channel members, lack of trust in collaboration and coordination, lack of IT experts, fear of failure, lack of direct incentives, Poor culture and structure of organization in GSCM, Lack of suitable performance management system, Lack of functional linkages between indicators- economic, social & ecological, Low ability of one channel member to influence another SC stakeholder, Lack of research and development on new materials and chemicals, Heavy investment in information technology and software, Poor IT human resource, Lack of integration of technology/compatibility, lack of government support and subsidy, and higher interest rate on loan, scarcity of green suppliers in Pakistan. These all are the significant barriers, which we have already discussed in Chaps. 5 and 6 in detail.

Many Pakistani firms are faced with financial barriers for starting green projects and implementing green practices in their firm's operations. There is no doubt that green supply chain management requires substantial fixed cost for the starting setup of smooth green operations such that firms need to purchase advanced technology for minimization of emissions and pollution associated with waste because modern high-tech is much more efficient and sophisticated than old technology. Besides, recycling, refilling, refurbishment, and remanufacturing activities also required a flexible design of products, which is only possible with high-tech usage in supply chain management operations. Many firms fear to implement green supply chain practices due to the massive fixed investment required in the starting phase of projects. On the other hand, several firm senior managers and directors confirmed that after the implementation of green supply chain practices in their firms' operations, the variable cost of end-to-end SC is amazingly lower than traditional SC operations because traditional SC is mainly based on polluting practices, which not only increase the consumption of chemicals and raw materials but also reduce firms' environmental performance with the negative image of firms.

Due to very tough competition and uncertainty in the domestic and international markets, it is challenging for the firms to keep the cost lower and to adopt green supply chain management practices at the same time. That is why competition plays a critical role in the implementation and non-adoption of green practices. Lack of awareness of green products and services in society is reflected as barriers in adopting environment awareness practices. The lack of knowledge in society means that customers are still not conscious of green products and their benefits to society. Lack of knowledge and understanding among suppliers acts as significant hurdles for the adoption of green supply chain management practices. The lack of understanding among supply chain stakeholders, poor supplier commitment/reluctant to change towards GSCM, and unwilling to exchange information with supply chain partners are the main barriers related to suppliers. Higher quality of human resources, such as having better training or education, will support and help the enterprises in the adoption of green supply chain management. Quality human resources can provide new ideas for companies, share knowledge, and use new technologies to solve the problem. However, due to the economic barrier, the quality of human resources becomes a barrier. Therefore the poor quality of human resources is a critical and significant barrier to adopt green supply chain management in manufacturing firms of Pakistan (Ref. Chap. 5).

Training and education on the topic of environmental-friendly policies and initiatives are the prime desires for attaining successful adoption of green supply chain management practices in manufacturing firms of Pakistan. Management may encourage employees to take training on green policies and legislation. Companies may also provide rewards in terms of cash and appreciation certificates for employees. We find that lack of training on green supply chain management practices and environmentally friendly concepts is one of the key barriers faced by manufacturing firms of Pakistan.

The lack of senior management commitment is a crucial barrier to adopt GSCM in manufacturing firms of Pakistan. Senior management commitment is very critical in the success of green projects and supply chain partners, aiming to adopt sustainable practices. The significance of top management beliefs and commitment has been highlighted in the findings. Hence, without proper monitoring and support from senior management, it is almost impossible to adopt green supply chain management practices successfully. Management should be proactive rather than reactive in going green in every possible aspect.

6.2.2 The Key Drivers

The findings reveal some key drivers of green supply chain projects, which increase the pressure on the firms to adopt green practices in their business operations for environmental sustainability in the long term and also for the enhancement in financial performance. The critical internal motivating factors are recycling facility of the firm; stakeholder pressure; economic benefits; profitable business opportunities; managerial realization of environmental degradation, protecting a business from risks of environmental factors, quality of products, lower rejection, and defect products; warranties and claim policies; environment protection system certification; benchmarking systems; cross-functional team; employee involvement; supplier commitment; cooperation with suppliers and customers; employee pressure; early involvement of supplier in green product development; more significant export opportunities; international trade unions; media pressure; organizational performance; financial performance; increase overall customer loyalty level; image building; and increased market share.

Government environmental friendly policies act as a significant driver that promotes firms to limit the use of nonrenewable resources and to reduce GHG (greenhouse gas emissions). Environmental policies targeting directly emissions are the most important single element of a cost-effective environmental policy strategy. Laws and legislation increase the threats of fines for noncompliance among firms. Global environmental requirements have shifted from pollution controls at manufacturing plants to the actual life cycle of the goods. This increases the scope of involvement by governments to involve green SC (supply chain) initiatives. Customer demand for green products is found as one of the most critical drivers for green initiatives. Customers are becoming increasingly aware of the negative impact caused by certain business activities on the environment. The customers and the public exert pressure on firms to adopt green practices in their supply chain operations, ultimately minimizing the harm these entities cause on the environment. Customers take the environment into concern when making to buy or make decisions. Customers demand environmentally friendly products by asking for ISO 14001 certification.

Environmental friendly image/green image and reputation as a driver of green supply chain management that offers a "greening image" where a manufacturing product is to be used. More critically, green image and reputation refer to the positive recognition by customers of firms that involve production processes that are greened. The green supply chain management adoption may enhance both the reputation and image of the company simultaneously. A firm will not only achieve a competitive advantage by attracting customers through a positive green image but also is eligible for advantages such as more export opportunities, reducing goods inspection time on customs clearance. Based on the above, it can be concluded that the projection of the positive green image can lead to higher profitability and reduction in total cost, which will, in turn, increase the competitive edge of a firm.

Social responsibility has become a critical factor in creating a global competitive advantage. Firms are developing green products to fulfill the customers' needs and requirements in order to maintain competitiveness in the domestic and international markets. The deterioration of environmental sustainability in the last couple of decades has increased public awareness of environmental issues. Society has demanded products from the firm, which has a greener operation. The public and media pressure causes most firms to review their manufacturing processes and practices within their firm's operation. The disruption caused by the environmental awareness of the public could generate opportunities for firms to grasp new customers who want environmentally friendly goods.

On the other hand, the researcher also finds additional motivating factors, which influence firms to adopt green/environmental-friendly practices in their supply chain and business operations. The key motivating factors include climate change, government rules and regulation, customer pressures, investor pressure, strict supervision by regulatory authority/governmental staff, pressure from media, pollution control air emissions, pressure from NGOs, special tax exemption for ISO-certified firms, ease in getting loans from financial institutions/banks, new market opportunities, reducing risks of consumer criticism, and financial incentives from international communities. These all are the key motivating factors, which enhance the firms' performance in the long run.

Due to this research, the key conclusions have been drawn from this research study.

The green supply chain management practices are continuously growing in the Pakistani manufacturing firms, while firms are implementing green techniques to enhance environmental sustainability and also increase their organizational performance. Besides, governments and regulatory authorities are promoting green practices in business operations. Furthermore, in the last couple of years, the government also enforced heavy taxes on polluted materials and chemicals.

In the last couple of years, the Pakistan government has strict its environmental laws to reduce environmental degradations. Governments enforce firms to conduct CSR (corporate social responsibility) activities to help society building such as greening the area of firms' plant/factory. Besides, polluted products and supply chains become a nontariff barrier because most countries' governments do not allow polluted products to enter their countries. Furthermore, polluted products are rejected all around the globe due to environmental degradation.

Customers' awareness is continuously increasing due to governmental and different NGOs' informational programs about environmental sustainability. Green supply chain specialists and professionals are fewer and fewer, which creates difficulties for firms to find the right professionals for their green project. Green suppliers' market in Pakistan is still smaller as compared to European countries, but since the last couple of years, hundreds of new suppliers entered the market to serve green manufacturing firms. Further, due to the latest/advanced technology, the adoption of green supply chain management practices is more comfortable as compared to before.

Consumers' awareness creates heavy pressure on firms to adopt green practices in their firms. Because the fundamental purpose of many firms is to earn a profit, and consumers are the ultimate buyers, fulfilling the consumers' requirements is making a profit.

Many firms adopt green techniques in business operations to enhance their financial performance and create a competitive advantage in their market. Early supplier involvement in product development also helps firms to adopt proper green supply chain techniques. Green purchasing is a technique of green supply chain. Green purchasing is creating a positive image of firms in the market. Besides, green purchasing also enhances financial performance with environmental performance.

In Pakistan, most firms are private and/or semiprivate, which is the biggest reason for fierce competition towards the adoption of green supply chain practices.

Undeniably, green firms' products are internationally acceptable. Most of the time, the government offers incentives. Firms also adopt green practices in their business operations to expand their customers' markets.

Green supply chain projects require proper monitoring to maintain better information sharing and coordination in the system. With the non-seriousness of senior management, many green supply chain projects will operate inefficiently. Undeniably, the top management role is vital during the adoption of green techniques in end-to-end supply chain management. Ecological design products not only enhance environmental sustainability but are also accessible to recycling, remanufacturing, and reusing.

Cooperation and coordination among supply chain members play an essential role in developing trust, which also supports and helps firms to adopt green practices successfully. Undeniably, the green distribution role is significant for controlling environmental degradation and increases customers' awareness of green product benefits.

Firms adopt new technology in their supply chain systems for enhancing green techniques/practices. Nevertheless, most of the time, new technology may create troubles for firms and their supply chain members in terms of "technology compatibility." So at the end of the results, technology becomes only a drain on firm profitability. Furthermore, in the context of Pakistan, green supply chain projects, and nongreen supply chain projects, banks offer loans on the same interest rate, which is somehow discouraging green projects.

Due to a lack of government support, Pakistani manufacturing firms are facing barriers and difficulties in the adoption of green practices. However, since the last couple of years, these barriers are not very significant due to the seriousness of federal and provincial governments towards environmental sustainability.

Green products provide more significant export opportunities for firms to increase their market share and size. Companies may improve their image and reputation in the domestic and international arena due to the implementation of green supply chain practices. Besides, firms grasp consumers' sympathy due to the adoption of green supply chain practices in their business operations.

International trade unions and nongovernment organizations create pressure on firms to adopt green practices and reduce their operations' harmful effects on environmental sustainability. Also, investors insist on organizations to adopt green practices. On the other hand, due to the adoption of green techniques, firms also are reducing consumer criticism and reshaping their market.

6.3 Research Contribution

This research objective was to identify key barriers and drivers faced during the adoption of green supply chain practices from the perspective of Pakistan, which we have discussed in Chaps. 4, 5, and 6 in detail. Furthermore, the researcher has highlighted some critical innovation points in the following paragraphs, while the supporting details on statistical and mathematical analysis can be seen in Chaps. 4, 5, and 6.

Many firms are facing financial constraints during the adoption of green practices due to expensive green materials. Besides, the government is also not providing incentives, subsidies, and tax exemptions on green materials, which create double burdens on manufacturing firms of Pakistan. On the other hand, if companies are adopting green supply chain practices, including ecological design and green techniques, the reason is to expand their market size and also grasp more significant export opportunities around the globe, especially in European countries.

Green supplier is fewer in Pakistan. Some manufacturing firms are importing green chemicals, materials, and components, which also incur extra costs into systems. On the other hand, in Pakistan, banks are providing loans to green and non-green projects on the same interest rate, which is significantly discouraging companies and senior management from adopting green supply chain practices in their firms' operation.

The researcher develops a model to test the impact of green supply chain practices on firms' performance from the perspective of Pakistan. There is no doubt that this is the first in detailed research conducted on manufacturing firms of Pakistan to identify the key drivers and barriers and the effect of green supply chain practices on manufacturing firms of Pakistan. This research proves that green practices in the supply chain build a competitive edge for manufacturing firms. Besides, the competitive edge also plays a significant role in multiplying the profitability of firms in the long run. In Pakistan, electronic and print media have a mighty influence on reshaping consumers' and inventors' minds. On the other hand, firms adopt green practices in their supply chain to increase products price because, after converting the polluted supply chain to the green supply chain, firms have moral rights to increase products price.

The difference in the understanding of the term "green supply chain management practices." Some supply chain members adopt green practices only for enhancing economic performance. On the other hand, firms are also facing a couple of crucial issues during the implementation of green supply chain practices in their firms' operations such as low ability of one channel member to control decision of another member in their supply chain and lack of information about green performance of suppliers; due to lack of collaboration with suppliers, supply chain partners start suffering from the problem of trust deficit. Finally, greater administrative and fixed cost with financial tax burden on the Pakistani manufacturing firms are also found to be a significant barrier.

Green information systems and senior management commitment play a direct and significant role in enhancing green practices efficiency and effectiveness. Besides, manufacturing firms also adopt green practices in their supply chain operations for increasing their products' quality and eliminating the waste from the end-to-end systems. Furthermore, Pakistani manufacturing firms are adopting green practices in their supply chain and business operations to increase customers' loyalty levels and increase repeat buying sales.

6.3.1 Achievements of Research Objectives

After the successful completion of the analysis, we found several barriers and drivers in the context of Pakistani manufacturing firms, which also has been discussed in Chap. 5. Further, we found that green practices (ecological design, green distribution, green transportation, and green manufacturing) have a positive and significant effect on Pakistani manufacturing firms' performance. The green supply chain is not only protecting natural beauty but also building a competitive edge and enhancing economic performance. There is no doubt that this is the first detailed research conducted on manufacturing firms of Pakistan to identify the key drivers and barriers and the effect of green supply chain practices on manufacturing firms of Pakistan.

6.3.2 Contribution to Knowledge (the Model)

GSCM research is newly conducted in the manufacturing industries of Pakistan, which significantly contributes knowledge in the literature review in terms of new barriers and drivers. Pakistan is a developing country, but people are more aware of environmental sustainability compared to other developing countries such as India, Bangladesh, and the Maldives. Due to consumers' and employees' pressure, firms are adopting a green supply chain in their business operations to sustain their profitability and existence in the market. This research will provide in-depth knowledge of GSCM practices in Pakistani manufacturing industries to supply chain professionals.

6.3.3 Find Out Reasons Why Practitioners Adopt or Not Adopt Green Practices

Undeniably, green supply chain practices are enhancing and positively correlated with firms' performance, but before implementation of green projects, companies should cater their capability in terms of financial, green knowledge of employees, availability of green suppliers for disturbance-free supply chain, availability of value-added technologies, security risks, unwillingness to collaborate with supply chain partners, and ethical values of firms. Due to the lack of capability, green projects can easily be converted in the drain on firm profitability without any betterment in environmental sustainability.

The corporate sector should adopt green supply chain practices in its business operations to improve firms' performance. Green supply chain practices build numerous fruitful benefits to firms such as cost reduction in terms of recycling, value stream mapping, reduction in resource utilization, building a positive image of firms in the domestic and international market, reducing risks of consumer criticism, increasing customers' trust and satisfaction, and reduction in packaging materials.

6.3.4 Implications for Practitioners

This research may be beneficial for the Pakistani manufacturing industry towards a better understanding and management of economic and societal issues in the supply chain. Our empirical results provide a theoretical framework by identifying six critical success factors (internal environmental management, green information systems, ecological design, green distribution, and green transportation) and their relationships with performance outcomes. This paper suggests that green practices (ED, GD, GT, IEM, GIS, etc.) have a significant impact on GSCM initiatives and performance outcomes. The critical success factor identified in the present research will help to achieve sustainable supply chain practices in the Pakistani manufacturing industry, which will satisfy regulatory or compliance requirements and at the same time will also help to achieve economic, environmental, and organizational performance. This paper may help supply chain managers and practitioners to understand various GSCM implementation issues and how they can be used gainfully to improve their practices and performances in the supply chain towards sustainability development.

6.3.5 Implications to Government

The present research may help regulatory bodies and policymakers to develop healthy regulatory norms/policies to promote green practices (ED, GIS, etc.) towards sustainability in the Pakistani manufacturing industry context. The Pakistan government also needs to strict its environmental laws for better environmental sustainability, while the government needs to provide more incentives and subsidies with lower interest rate loans for green projects. This research is to persuade stakeholders/regulatory bodies to pay attention to the impacts of green practices and expected organizational performance outcomes, which may help Pakistani manufacturing industry or other emerging economies' ineffective implementation of GSCM towards sustainability. Finally, the government needs to protect green projects and firms in terms of providing loans on the lowest interest rate, subsidies on green materials, and tax exemptions for green projects to encourage and motivate firms to adopt green supply chains in their business operations.

6.4 Research Recommendations, Limitations, and Future Research

In order to improve green supply chain practices in manufacturing organizations of Pakistan, senior management and employees should be well-committed towards green projects because most green projects failed due to a lack of senior management support. On the other hand, top management commitment also enhances employees' motivation. The green supply chain is a complete ideology and/or process. So firms should provide proper training to their employees for significant achievement.

Supply chain members (downstream and upstream) should coordinate and collaborate with the nucleus firm for channel performance optimization because green projects mainly require more magnificent information sharing with upstream and downstream. On the other hand, before starting the green project, project feasibility report, return on investment, and cost-benefit analysis should be on a realistic basis to avoid any misunderstanding because most green projects have been failed due to underestimation and/or overestimation. Besides, supply chain stakeholders should perceive the same definition of green supply chain management, because, in this research, we also found that some firms are only adopting green practices for enhancing their financial performance without catering to environmental sustainability. Finally, firms should hire proper consultants and/or specialists for monitoring green projects. Also, the supply chain specialists should be the qualified persons of CSCP (certified supply chain professionals). Furthermore, several firms adopt green practices only to enhance their financial performance, which loses the concentration of firms on environmental sustainability and the essence of green sustainability projects.

Usually, every research has its limitations, and "nothing is perfect," and each research has its aim to identify valid and reliable results. So covering all aspects of the topic is not possible for the researcher practically. Because of financial and human resource constraints, the research may have some limitations such as researcher collecting the data from only manufacturing firms of Pakistan, data collected from primary sources including, questionnaires and interviews, this research adopting PLS-SEM Modeling techniques to test hypothesis, that research is cross-sectional, and that all firms are privately owned firms in study sample size. Therefore, in the future, more researchers can collect secondary data and conduct a study to explore further barriers and drivers at the macro-level. Further, some other business performance (quality, technological, advancements, adaptability, flexibility, and delivery) can also be investigated. The sample drawn was focused on some particular section (Pakistani manufacturing firms); thus in future studies, a variety of sectors (electronics and electrical, power, construction, chemical, plastic, etc.) and a more geographically diversified sample of organizations can be used. The findings obtained from other sectors can also be compared with the results of this study.

The future researchers may also research other than the manufacturing industry with some new quantitative indicators such as government policies, ISO-certified firms for environmental-friendly, cooperation with suppliers and stakeholders, and green purchasing. However, future researchers should also include social aspect indicators from secondary databases such that health diseases are negatively correlated with green practices and green supply chain management. Besides, green supply chain and tourism, foreign direct investment inflows, gross domestic products, and industry value added can be positively correlated with the perspective of Pakistan.

Appendix A: Correlational Matrix

Correlational Matrix Item Wise

S. A. R. Khan, *The Critical Success Factors of Green Supply Chain Management in Emerging Economies*, EAI/Springer Innovations in Communication and Computing,
https://doi.org/10.1007/978-3-030-42742-9

	GIS1	GIS2	GIS3	GD1	GD2	GD3	GD4	GT1	GT2	GT3	GT4	ED1	ED2	ED3	GM1	GM2	GM3
GIS1	1																
GIS2	0.430[a]	1															
GIS3	0.410[a]	0.641[a]	1														
GD1	0.242[a]	0.420[a]	0.395[a]	1													
GD2	0.275[a]	0.506[a]	0.442[a]	0.596[a]	1												
GD3	0.348[a]	0.468[a]	0.367[a]	0.191[a]	0.299[a]	1											
GD4	0.243[a]	0.503[a]	0.442[a]	0.459[a]	0.453[a]	0.258[a]	1										
GT1	0.187[a]	0.352[a]	0.319[a]	0.110[b]	0.191[a]	0.406[a]	0.156[a]	1									
GT2	0.361[a]	0.446[a]	0.453[a]	0.214[a]	0.338[a]	0.491[a]	0.333[a]	0.455[a]	1								
GT3	0.232[a]	0.464[a]	0.494[a]	0.640[a]	0.585[a]	0.270[a]	0.634[a]	0.148[a]	0.326[a]	1							
GT4	0.072	0.075	0.015	−0.049	0.028	0.076	0.075	0.220[a]	0.115[b]	−0.013	1						
ED1	0.358[a]	0.491[a]	0.412[a]	0.238[a]	0.390[a]	0.522[a]	0.356[a]	0.448[a]	0.540[a]	0.344[a]	0.083	1					
ED2	0.393[a]	0.480[a]	0.398[a]	0.234[a]	0.372[a]	0.726[a]	0.303[a]	0.556[a]	0.666[a]	0.327[a]	0.091	0.694[a]	1				
ED3	0.478[a]	0.389[a]	0.360[a]	0.265[a]	0.316[a]	0.381[a]	0.208[a]	0.243[a]	0.303[a]	0.243[a]	0.089	0.317[a]	0.362[a]	1			
GM1	0.203[a]	0.316[a]	0.285[a]	0.215[a]	0.283[a]	0.292[a]	0.226[a]	0.249[a]	0.298[a]	0.184[a]	0.071	0.256[a]	0.293[a]	0.293[a]	1		
GM2	0.425[a]	0.536[a]	0.455[a]	0.259[a]	0.345[a]	0.410[a]	0.368[a]	0.281[a]	0.450[a]	0.312[a]	0.003	0.458[a]	0.482[a]	0.428[a]	0.279[a]	1	
GM3	0.510[a]	0.472[a]	0.382[a]	0.307[a]	0.326[a]	0.416[a]	0.284[a]	0.271[a]	0.446[a]	0.257[a]	0.057	0.403[a]	0.498[a]	0.742[a]	0.428[a]	0.714[a]	1
ECP1	0.562[a]	0.473[a]	0.389[a]	0.246[a]	0.317[a]	0.323[a]	0.353[a]	0.165[a]	0.369[a]	0.294[a]	−0.003	0.350[a]	0.428[a]	0.434[a]	0.207[a]	0.480[a]	0.542[a]
ECP2	0.517[a]	0.473[a]	0.380[a]	0.256[a]	0.303[a]	0.336[a]	0.263[a]	0.221[a]	0.331[a]	0.226[a]	0.031	0.437[a]	0.439[a]	0.413[a]	0.171[a]	0.520[a]	0.496[a]
CA1	0.266[a]	0.489[a]	0.351[a]	0.254[a]	0.368[a]	0.376[a]	0.349[a]	0.302[a]	0.335[a]	0.322[a]	0.064	0.500[a]	0.369[a]	0.238[a]	0.187[a]	0.420[a]	0.318[a]
CA2	0.384[a]	0.550[a]	0.466[a]	0.383[a]	0.500[a]	0.384[a]	0.379[a]	0.314[a]	0.430[a]	0.423[a]	0.065	0.484[a]	0.421[a]	0.357[a]	0.296[a]	0.457[a]	0.423[a]
CA3	0.336[a]	0.535[a]	0.422[a]	0.327[a]	0.456[a]	0.449[a]	0.353[a]	0.319[a]	0.463[a]	0.347[a]	0.099	0.522[a]	0.478[a]	0.349[a]	0.305[a]	0.420[a]	0.386[a]
ORP1	0.439[a]	0.574[a]	0.453[a]	0.350[a]	0.474[a]	0.430[a]	0.397[a]	0.268[a]	0.422[a]	0.384[a]	0.019	0.478[a]	0.460[a]	0.386[a]	0.303[a]	0.512[a]	0.457[a]
ORP2	0.509[a]	0.585[a]	0.501[a]	0.252[a]	0.392[a]	0.457[a]	0.422[a]	0.319[a]	0.501[a]	0.342[a]	0.054	0.544[a]	0.509[a]	0.436[a]	0.252[a]	0.480[a]	0.479[a]
ORP3	0.395[a]	0.464[a]	0.468[a]	0.368[a]	0.341[a]	0.400[a]	0.332[a]	0.316[a]	0.466[a]	0.352[a]	0.087	0.479[a]	0.499[a]	0.410[a]	0.241[a]	0.432[a]	0.426[a]
IEM1	0.467[a]	0.363[a]	0.351[a]	0.254[a]	0.268[a]	0.365[a]	0.256[a]	0.322[a]	0.344[a]	0.259[a]	0.069	0.335[a]	0.396[a]	0.353[a]	0.190[a]	0.425[a]	0.416[a]

IEM2	0.542[a]	0.542[a]	0.419[a]	0.280[a]	0.377[a]	0.425[a]	0.388[a]	0.251[a]	0.448[a]	0.340[a]	0.039	0.438[a]	0.518[a]	0.409[a]	0.281[a]	0.519[a]	0.563[a]
IEM3	0.498[a]	0.565[a]	0.478[a]	0.219[a]	0.370[a]	0.430[a]	0.419[a]	0.326[a]	0.504[a]	0.338[a]	0.045	0.525[a]	0.484[a]	0.423[a]	0.218[a]	0.470[a]	0.463[a]
IEM4	0.392[a]	0.460[a]	0.464[a]	0.365[a]	0.329[a]	0.375[a]	0.326[a]	0.306[a]	0.464[a]	0.352[a]	0.098	0.455[a]	0.482[a]	0.394[a]	0.226[a]	0.427[a]	0.414[a]
ENP1	0.226[a]	0.396[a]	0.361[a]	0.208[a]	0.312[a]	0.415[a]	0.194[a]	0.473[a]	0.502[a]	0.234[a]	0.088	0.452[a]	0.480[a]	0.294[a]	0.307[a]	0.363[a]	0.345[a]
ENP2	0.430[a]	0.549[a]	0.448[a]	0.336[a]	0.440[a]	0.400[a]	0.394[a]	0.246[a]	0.421[a]	0.374[a]	0.012	0.467[a]	0.437[a]	0.367[a]	0.273[a]	0.514[a]	0.454[a]
ENP3	0.499[a]	0.574[a]	0.491[a]	0.217[a]	0.368[a]	0.442[a]	0.405[a]	0.336[a]	0.515[a]	0.340[a]	0.058	0.536[a]	0.506[a]	0.415[a]	0.222[a]	0.482[a]	0.471[a]

[a]Correlation is significant at the 0.01 level (two-tailed)
[b]Correlation is significant at the 0.05 level (two-tailed)

Appendix B: Questionnaire

Descriptions
Variable 1: Internal environmental management
Top management commitment towards GSCM
Cross-functional cooperation for environmental improvement and environmental management systems
Total quality environmental management and ISO 14001 certifications
Environmental compliance and auditing programs
Variable 2: Green information systems
Monitoring emission and waste production
Improving decision-making by executives by highlighting sustainability problems
Reducing energy consumption and limiting carbon emissions
Descriptions
Variable 3: Green manufacturing
Green practices in manufacturing processes and reducing fossil fuel energy consumption
Using green materials in manufacturing
Decreasing toxic and hazardous chemicals in manufacturing processes
Variables 4: Eco-design
Design of products for reduced consumption of material/energy
Design of products for reuse, recycle, recovery of material, or components parts
Design of products to avoid or reduce use of hazardous products and their manufacturing process
Variables 5: Green distribution
Distribution centers are working on environmental sustainability
Distribution centers are nearby customers
Using the shortest route for arriving on destination
Variables 6: Green transportation
Reducing transportation cost

S. A. R. Khan, *The Critical Success Factors of Green Supply Chain Management in Emerging Economies*, EAI/Springer Innovations in Communication and Computing,
https://doi.org/10.1007/978-3-030-42742-9

Using fuel-efficient vehicles to reduce emissions
Vehicles maintenance
Variables 7: Environmental performance
Reduction of air emissions, effluent waste, and solid waste; decrease in consumption of hazardous/harmful, toxic materials
Improvement in an firm environmental situation
Variables 8: Competitive edge
Customers positive feedback on green products and repeat buying
Improved firm image and reputation
Variables 9: Economic performance
Decrease in cost for energy consumption and decrease in cost of material purchasing
The decrease in fee of waste treatment and waste discharge
Variables 10: Organizational performance
Improved capacity utilization, increase in product quality and amount of goods delivered on time
Avg. ROI and profit growth over the past 3 years
Avg. sales volume and in US$ growth over the past 3 years

Note: This questionnaire is used to collect the data from the manufacturing firms. The questionnaire is based on the Likert Scale (1 = strongly disagree and 7 = strongly agree)

Appendix C: Interview Questions

Mainly the unstructured interviews are based on the following questions:

- What are the main barriers faced by manufacturing firms during the adoption of GSCM practices?
- What are the key drivers and motivating factors involved behind the adoption of GSCM practices?

S. A. R. Khan, *The Critical Success Factors of Green Supply Chain Management in Emerging Economies*, EAI/Springer Innovations in Communication and Computing,
https://doi.org/10.1007/978-3-030-42742-9

Index

S. A. R. Khan, *The Critical Success Factors of Green Supply Chain Management in Emerging Economies*, EAI/Springer Innovations in Communication and Computing, https://doi.org/10.1007/978-3-030-42742-9

The manufacturer's authorised representative in the EU is Springer Nature Customer Service Centre GmbH, Europaplatz 3, 69115 Heidelberg, Germany. If you have any concerns regarding our products, please contact ProductSafety@springernature.com

Printed and bound by CPI Group (UK) Ltd, Croydon, CR0 4YY
30/09/2026
02225647-0007